BEDFORDSHIRE

Mills

by

Hugh Howes

design & drawings
by
Sarah Garner

County Planning Department

Thurleigh Tower Mill in 1935. The missing sail had been blown off in 1917.

Royal Commission on Historical Monuments (England). HES Simons Collection

Previous page

Perhaps looking less at home in Bedfordshire than in the 'wild west' is this predecessor of Stoke Mills, Sharnbrook. The mill house is little changed today. Based on a painting by C F Langley 1832.

This book is set on 10 on 11 Times Medium.
Typeset and printed by Watkiss Studios.
Biggleswade, Beds.

Graphics by the County Planning Dept.

Bedfordshire

ISBN 0 907041 06 X

Contents

Cranfield tower mill was built in the late 1840s. Seen here in its prime it had four patent sails which drove three pairs of stones and flour dressers. It worked until 1935 when one sail was wrecked in a tail wind. Cf photo on p.25.

Photo supplied by B W Anstee

Acknowledgements

I am particularly grateful to John Jordan whose family have been milling at Holme Mills, Biggleswade, for more than 130 years. He provided me with much information about milling on the river Ivel, including several incidents from the family's history and many illuminating details of the milling process.

William Hipwell too, gave me acccess to his Company's minutes which vividly explain the decline of local milling in the post War period. He also showed me his late father's articles about the mills of Sharnbrook, which I have quoted.

John Goodman, who still keeps Flitwick water mill working, gave me several insights into the problems of milling on a small river like the Flit.

John Clover has lived at Kempston Mill all his life. His father ran it in partnership with George Horne. He kindly made available to me many press cuttings and his own industrial records and demonstrated how to *dress* a millstone.

I was also lucky enough to have been able to talk to the late John Buckmaster Linney about Doolittle Mill, Totternhoe, only a matter of weeks before his death.

I am grateful too to Mrs. Lewis for showing me around Taylor's Mill at Stotfold, to Mr. James for explaining the history of the Simmons family at Leighton Buzzard, to Jim Battison of Blunham for his recollections of Blunham Mill more than 80 years ago, to Mr. L. W. Vass for telling me about the restoration that he carried out at Barton Mill and to Mr. B. W. Anstee concerning Cranfield Mill.

I have had access to a tape recording, made in 1975 by the pupils of Lincroft School, Oakley, of interviews with the late Mr. Filsell, the last miller of Keysoe, and with Mr. Keech, the restorer of Stevington Mill.

It was fortunate that most of the research into the history of windmills had already been completed by Peter Dolman when he was at Bedford Modern School. Furthermore, several technical descriptions had already been compiled by Kenneth Major. My thanks are due to both of them. I am also grateful to John Reynolds who made available the material from the survey of Woburn Abbey Mill by his students at the Architectural Association.

Kevin Fadden of the Ampthill Archeological and Historical Society kindly made the recording of Mr. Charles Scott available to me.

Many of the photographs came from the County Photographic Unit and I am grateful to Ken Whitbread and Dave Stubbs in this connection. The late Harry Meyer of Letchworth kindly loaned me his collection of photographs of Bedfordshire windmills which he took in the 1920s and '30s, many of which are in the book.

Fortunately there are photographic records of many mills but equally there are a number where the absence of a visual record must remain a tantalising blank. What, for instance, did the six-sailed windmill which stood at Leighton Buzzard until about 1890 look like; or the strange structure that stood between 1880 and 1903 at Hall End in Maulden parish? According to the *Victoria County History* it was like a barn with a turret at one end, on which the sails were supported but which could not be *winded* because of the nature of the structure. We do not even have a picture of Stevington Mill before its radical rebuild in 1921. If readers are spurred into finding interesting mill photographs in ancient family albums, the County Planning Officer would be delighted to hear from them.

I am particularly grateful to those who have worked on the text and offered criticism and correction — Peter Dolman, Kenneth Major and, of the County Planning Department, David Baker, Alan Cox, Stephen Coleman, John Wood and especially to Sarah Garner, Planning Technician, who not only provided the maps and drawn illustrations including the splendid cover but also took many constructive initiatives whilst setting out the individual pages.

Photographs have been kindly provided from many individuals and bodies and are reproduced with their permission. The source of these pictures is identified in the captions.

Finally I would like to thank Anne Ashmore for producing orderly typescripts from my disorderly preliminary manuscripts and to my wife Monica without whose laborious and painstaking work this book would be infinitely less readable!

Horse-drawn waggon of George Horn of **Kempston Mill** at the turn of the century.

Photo supplied by J. Clover

List of illustrations

Maps

NB *The numbers shown on the maps correspond to the reference numbers in the gazetteer which starts on page 85. Those mills whose sites are not known precisely have been omitted from the maps.*

Foreword

Windmills and watermills are a fascinating part of the Bedfordshire rural industrial heritage. Many of the working mills which existed fifty years ago have now disappeared, together with the way of life and kind of work they supported.

Hugh Howes, a member of my strategic planning team, has made an extensive private study of Bedfordshire mills. He has concentrated especially upon recollections of former millers and photographs of their buildings. I know that his work will be of great interest and I hope that it will encourage the conservation of windmills and watermills that survive.

A. M. Griffin

County Planning Officer

Preface

John Jordan told me, *"There were four hundred mills in Bedfordshire actually making flour in 1855."* Now his Holme Mill at Biggleswade is the one remaining that still manufactures flour in the county. Within a lifetime a traditional local industry has disappeared and with it many of its artefacts, including the mills themselves.

Where were these mills? How many have survived? Why have they stopped making flour? And what of the men who worked these mills?

Flour milling had started to decline before the First World War and had reached the point of virtual extinction by the 1960s. John Goodman, owner of one of the few surviving mills at Flitwick, explained what happened—

> "Up to 1914 they used to make flour here but the modern reduction processer roller machines made the old stone-ground flour obsolete."

Now Flitwick Mill, along with a handful of other surviving water mills, has switched to providing crushed barley for horse food. So this trend has all but destroyed a heritage of more than a thousand years of milling.

I have talked to a number of mill owners and local people. I have also had access to tape recordings of people who in their final years recorded the memories of milling in days gone by. I have drawn extensively on these interviews and transcribed many extracts in order to capture some of the flavour of the days when the mills were still producing flour.

It is this historical and social aspect on which I have concentrated rather than on the detailed technical workings of mills themselves. I have only gone into extensive technical detail where it is helpful in explaining the historical development of milling. Nevertheless those of technical disposition will find many of the appropriate details of individual mills in the last two chapters. Further information is available on files held in the County Planning Department's Sites and Monuments Record.

Facing page

Final flowering of Victorian millwright's art. Underneath the cap of **Biggleswade's tower mill** showing the massive iron brakewheel, brake and the curb on which the cap and sails rotated to face the wind.

Beds C. C. Photographic Unit

Overleaf

Primitive Forebear; A small open-trestle post mill with unusual vertical boarding. Its site, now a green on a golf course, is close to Woburn Sands but is within Aspley Guise Parish. The Photograph was taken in about 1900 shortly before the mill was pulled down. Part of one sail is missing and the mill must already have been out of use for some time.

Photo supplied by Peter Dolman

Chapter 1
Early Days

For a thousand years the fortunes of local mills and millers have fluctuated, but it is only during the last sixty years that they have declined so rapidly. The watermill had probably been introduced into this country by the Romans. We do not know much about milling in the area of Bedfordshire during these early times. A few stone querns, for hand milling, have been found. However, in late Saxon times Bedfordshire was prosperous and almost every manor had its mill to which the peasants had to take their corn to be ground. John Clover explained that,

> ". . . under the law of the milling *soke* only properly established mills were allowed—usually of ecclesiastical or manorial foundation."

A windmill would often be built in the midst of open fields whose tenants were obliged to use it exclusively. This was an essential part of feudal obligations. The Lords of the Manor held a milling monopoly and required tenants to bring their corn to ground in the manorial mills. Indeed by the time of the Domesday Survey there were 103 mills recorded in the county.[1] Many that survive today may still occupy their predecessors' ancient sites.

See you our little mill that clacks,
So busy by the brook?
She has ground her corn and paid her tax
Ever since Domesday Book.

Puck's Song—Rudyard Kipling

Sundon post mill with sail cloths furled. It was moved to a site at Tebworth Road, Toddington, in 1845 where it survived until blown down in the 1880s.

County Record Office

We cannot be certain that all Domesday mills were watermills; some may have been powered by hand or animals. A further qualification is that the customary definition of a mill is one pair of millstones. Two pairs within one building would be two mills and this may have affected the Domesday calculations. For example, William Hipwell explained,

> "It is absolutely wrong to talk about Stoke Mill. In the old days when they had mill stones a mill that had one pair of stones was a mill. If it had more than one pair of stones, and that mill had four, then it was Stoke Mills."

These 11th and 12th century mills must have been relatively primitive. A crude wheel drove a stone quern which would have been enclosed in a rudimentary hovel. A flat county like Bedfordshire might not be regarded as favourable for the widespread use of water power. Nevertheless almost any stream can

[1] **The Domesday Watermills of Bedfordshire** by William Austin, *Bedfordshire Historical Record Society*, Vol 3, 1916.

be dammed to provide a sufficient level of water to drive a wheel, if only for relatively short periods. Mr. L. W. Vass explained how this applied to Barton Mill.

> "It is an artificial bank now forming the mill head and I understand the soil for that came from when they made the Barton Cutting" (for the A6 road through a chalk ridge).

Even a hundred years ago the rivers were usually wider and had more copious flows than today when water tables have often been lowered by drainage works. Mr. Vass added,

> "There was a considerable quantity coming down, more than there is now. The volume is considerably decreased than what it was years ago because of extraction,"

The building of mills in such locations, few of which have survived, reflects economic circumstances very different from those of today. The cost of transport in those times put a premium on the *local* availability of milling facilities. However inadequate the source of power, their proximity was of paramount importance for the work of local farmers.

Thus watermills can be found not only on the main rivers, the Great Ouse, the Ivel, the Flit and the Ousel, but also on the minor tributaries whose volumes of water could be very meagre. Indeed Bedfordshire lies on the fringe of East Anglia, a region characterised by a flat landscape with unhurried streams. These could provide only limited power to turn waterwheels, although large wheels and lower gearing could get the millstone to turn at the right speed. Furthermore, the low rainfall of this region meant that

Barton Mill in 1905. Showing the mill head, the chimney serving a steam engine which operated the mill when water supply was not sufficient.

Photo supplied by L. W. Vass

streams were often small and their flow irregular. Yet this was also the main cereal-growing area of the country, and with the population rising steadily until at least the mid-13th century, arable farming prospered. This combination of circumstances must have encouraged the harnessing of wind as a new source of power to meet the demands for greater and more constant milling. Windmills were most numerous in East Anglia and Bedfordshire had its share of the 1,500 which were once dotted around this region.

Part of the Totternhoe Graffiti

Between the river valleys lie substantial tracts of countryside with no accessible water power. They comprise the rolling countryside of North Bedfordshire, the western part of Mid-Bedfordshire and the Chiltern scarp slope. There is no reference to any windmill in Domesday although they were already well known in the Low Counties at that time. They were not to become common over here for a further 150 years, when Bedfordshire was in the forefront of this new technological development. A windmill was recorded at Leighton Buzzard by 1212 and further ones elsewhere in the county in the 1220s and 1230s. They must have been amongst many built locally during the Middle Ages, and known only from passing references in documents. They were mostly light, simple structures that could be easily erected, moved and demolished leaving little trace of their existence.

Even if few of Bedfordshire's windmills have survived, the county does posses a unique piece of evidence of what the late mediaeval windmill looked like. On the soft walls of Totternhoe Church, which are built of local *clunch* (a hardened form of chalk) are numerous graffiti of windmills.[2] These are accompanied by the dates 1748 and 1753. They perhaps represent the primitive post mill which stood at Totternhoe between the church and Doolittle Mill. Its windshaft and spur wheel are believed subsequently to have been used as the top half of the upright shaft at Doolittle Mill. The graffiti show open trestle mills with a simple gabled buck and short square sails and a long tailpole characteristic of the late mediaeval period. It was only during the 19th century that open trestles were enclosed by a round house.

Looking up inside **Doolittle Mill, Totternhoe.** It was built between 1815 and 1825 but the great spur wheel of the windmill, shown here, is much older and is believed to have come from an earlier mill.

Beds. C.C. Photographic Unit

In milling, as elsewhere, the experience of the bubonic plague (Black Death) in the 14th century marked the end of an era. The period of agricultural expansion was over, the population had been nearly halved and the arable cultivation of much land ceased. The number of water and windmills in use must have fallen as a result. They would not have been replaced in the later mediaeval period because so much arable land had been permanently converted into pasture, thus reducing demand. There are records of windmills in over eighty parishes in Bedfordshire, and references to 141 windmills which ground corn, including Tilbrook and Eaton Socon, then within the County but now in Cambridgeshire.

There was also a *pug* mill at Husborne Crawley which was built at the turn of the 18th and 19th centuries to provide power for

2 **Windmill Graffiti at Saint Giles' Church, Totternhoe,** Terence Paul Smith. *Bedfordshire Archaeological Journal,* Vol 14, 1980.

A scene in about 1820 by Thomas Fisher showing the rare horizontal 'pug' mill which supplied clay to the Crawley Kiln (centre) and the Husborne Crawley watermill (right).

Photo supplied by North Beds. Borough Council

chopping up and blending clay prior to moulding as part of the brick-making process. It was an unusual design known as a *horizontal* windmill. The sails consisted of a number of pivoted shutters set in a horizontally revolving cylinder so as to catch the wind on one side only. One side flapped open to allow the wind to blow past while the opposite shutter blew closed to present a sail area to the wind. This arrangement was attached to a vertical shaft which emerged through the apex of a conical roof.

Like most horizontal mills it appears to have been short-lived. A letter of January 1832 to the Duke of Bedford, who was keen on applying technological novelty on his estates, concerns repairs and improvements to various parts of the estate property. One part is headed *Crawley Kiln* and says

> "to erect machinery to the Windmill for the purpose of turning the pug mill."

The cost was £30. The mill was probably demolished in 1840 and definitely had gone by 1884 when the 6-inch Ordnance Survey map was published.

Post mills might be physically moved to new sites some distance away. One at Sundon was moved to Toddington in 1847 where it was placed on an old-established mill site. Similarly a post mill known as Flemmon's mill stood on the Heath Road on Shenley Hill at Leighton Buzzard until 1860 when it was dismantled and re-erected at Whipsnade, where a round house was built for it. Thomas Twiddell had this move carried out by a millwright called Anderson from Totternhoe, for the benefit of his son. Young Mr. Twiddell cannot have made much of a success of it for it was demolished in 1877.[3]

[3] *Luton Post-Echo*, 8th October 1981

The period 1750-1850 has been described as the golden age of milling and most mills were substantially rebuilt during the century to accommodate the latest industrial technology. There was a trend back from cattle farming towards arable, reviving the demand for milling. In the last half of the century many of the agricultural improvements associated with such famous names as *Turnip* Townshend and Robert Bakewell were taking place, and milling technology made parallel advances. The windmills at Bolnhurst, Keysoe, Riseley and Stevington date from this time. So do most of the Upper Ouse watermills together with Hyde Mill on the Lea and the Flit mills at Flitwick, Maulden and Clophill. They were built in what industrial archaeologists call the *rural vernacular* tradition.

Some idea of Bedfordshire's mills in the early 19th century can be gained from notices advertising their sale. The roseate language of estate agents has changed little down the years.

Barton windmill, already at least 20 years old, was described in 1851 as —

> "the newly erected Post Windmill, with two spring sails and two cloth sails; driving one pair of 4 ft. 6 inch French Stones, now and for many years known as Barton Mills."

Holme Mills, Biggleswade, was described in best estate agents prose in 1829 as—

BEDFORDSHIRE

Capital Freehold Water Corn Mill
And most eligible situation
HOLME MILL, near BIGGLESWADE, Beds
To be SOLD by AUCTION
by Mr Hagger
At the Sun Inn and Hotel, Biggleswade, on Monday March 30th 1829 at Two o'clock in the Afternoon,
ALL that capital and complete WATER CORN MILL called Holme Mill situate upon the River Ivel, near Biggleswade, Beds, possessing an excellent stream of Water, two water Wheels, 16 Feet in Diameter (one seven feet wide, the other four and a half). Four Pair of Stones, Machines etc; with a substantial convenient, brick and attached messuage or Dwelling House; containing two Parlours, Kitchen, four Bedchambers, five Attics, brew house, with Servants' Room over it; excellent stables for cart and saddle Horses, Chaise-horse, Barn, Wagonshed, Piggeries, Gardens, planted with choice fruit Trees; Paddock, fish pond and other suitable Conveniences. Also EIGHT ACRES of luxuriant MEADOW LAND, adjoining the Premises; the Whole being Freehold and tithe free. The Premises are most delightfully situate a very short Distance from the Road from Biggleswade to Shefford

and Ampthill. Exceedingly well calculated for a Paper Mill, or to carry on an extensive Business as Corn Mill; is close to the Navigation from Shefford and Biggleswade to Lynn, and which Canal is expected at no very distant Period to communicate with the Grand Junction (if not carried to Ware) and thereby communicated immediately with the Metropolis forming one of the most desirable as well as delightful situations in the Kingdom; and a Purchaser may be accommodated with from £3,000 to £4,000 on Security of the Estate, if desired.

Holme Mill is situate 44 Miles from London, Biggleswade one, Hitchin 10, St Neots and Ampthill 12, Bedford 11, and Potton and Shefford four

For Further Particulars, and to treat for the same by private Contract, apply to Mr **LAXTON** on the Estate; or to Mr **CHAPMAN**, Solicitor, Biggleswade.

"Overshot" waterwheel at **Flitwick Mill.**
Beds. C.C. Photographic Unit

Chapter 2
Victorian Times

The 19th century brought fundamental changes which ultimately were to prove fatal for small-scale milling by the local man. Initially, the increase in grain production stimulated milling in the county. But by mid-century the towns, first Luton and later Bedford, were attracting the rural population and growing rapidly. There followed a rising demand for the bulk supply of flour to feed the urban masses and a corresponding reduction in demand for locally produced flour in the villages. John Clover explained that,

> "In the 1840's these county millers could make no end of money."

New measures were needed. W. H. Rogers' massive steam mill was built in the St. John's area of Bedford in 1885.

Because of the limited flow of most Bedfordshire rivers and the uncertainty of the wind the small local mills could not cope with the increasing demands placed on them. Many millers therefore improvised. The late Frank Hipwell recalled that the windmill at Sharnbrook was built,

> "to act as a standby or assistant to the *Old Mill* driven by the brook, in the hope that wind might be available when water was short. The juxtaposition of two such mills under the same ownership was not an unusual arrangement, giving the miller an alternative of two natural elements from which to draw power."

Auxiliary power at **Doolittle Mill, Totternhoe.** The wheel could take a belt drive from a steam engine used in later days to supplement the waterwheel.

Beds C.C. Photographic Unit.

But he also demonstrated the inherent inability of the old mills to respond to even more pressing needs,

> "Water sometimes *ran out* before his grist, when the unfortunate customer had to wait a longer or shorter period, depending on the flow of the brook at the time, for the balance of his meal to be made. When this happened the wheel-gate was shut down until sufficient water had collected to allow operations to begin again. However, time did not press a century ago!"

Time did soon begin to press. Increasingly millers adapted their mills to be driven by portable steam engines, traction engines and subsequently internal combustion engines or electric motors to overcome the problem of a reliable source of power at a critical time of year. The inventory of a lease in 1907 of Barton Mill includes a reference to *4 ft iron drum with crank complete to connect steam engine to the machinery in the mill.* So there followed a rapid decline of both windmills and watermills in Bedfordshire, as in the rest of the country.

There were three mortal blows between 1870 and 1880. A series of bad harvests in 1875, 1876 and 1877 made it even more difficult for small local mills to meet urban needs. Secondly, in 1875 Free Trade was introduced and wheat began to flood in from the United States, transported by the new steam ships and railways. These two developments resulted in a third, the building of very large steam-driven flour mills close to the ports. Using roller crushers instead of stones and benefiting from a more constant source of power they were far more efficient than traditional local mills and better able to supply bulk white flour for the expanding towns and cities.

The mills in the Ivel Valley successfully resisted the long decline for several years after most other mills in the county had succumbed. This has always been a prosperous agricultural area, with soil of excellent quality fully exploited down the centuries, and grain, especially barley, provided plenty of business for local millers. Only recently has its major town, Biggleswade, declined in importance: two watermills continued to operate into the 20th century.

The excellent road, water and rail communications were a further factor which made for the survival of the Ivel mills after decline had set in elsewhere. The Great North Road was an asset denied to other parts of the county. Then, in 1757, the Ivel Navigation was opened; it was extended to Shefford in 1842. Not only could the millers benefit from cheap transport but they were also entitled to levy a toll of three pence per barge passing through the locks

A Victorian "Industrial" style mill building at **Langford Mills** on the River Ivel.

Beds. C.C. Photographic Unit

adjacent to their mills as a form of compensation for loss of water from their head races[1]. In practice this was seldom a problem. Mr. Jordan said,

> "That was because you had plenty of water power. It is a very good river, the flow is always good, even in the summer."

The Great Northern Railway took away traffic from the Navigation which was sold off in 1876. This development too was favourable to the millers because of the access to new markets provided by the railways. Indeed it was even worthwhile to modernise these mills still further, dispensing with wheels and stones and substituting turbines and roller crushers. So it was on a sound economic basis that the Ivel millers tore down the old mills and from about 1850 onwards started erecting the large *industrial* mill buildings, most of which have survived today.

At the end of the 19th century two new technological developments were introduced, the water turbine and the roller mill. They were to revolutionise milling. They rendered obsolete all the windmills and forced water mills either to modernise or to go out of business.

Turbines were installed during this period at Stoke Mills, Holme Mills, Langford Mill, Bellowe's Mill, Leighton Buzzard, and Kempston Mills. John Clover remembered,

> "the turbine was a great asset. We used to get 60-70 horse power from it."

And John Jordon said,

> "We have a 30 hp turbine running. It was put in by Gilbert and Gilkes of Kendal. It is an interesting old turbine. The rotor is about 10 feet diameter and it has a six-inch shaft up the middle. It all turns round on a piece of wood at the bottom. It is *lignum vitae*, the hardest wood in the world. It is lubricated by water. I thought we would lift up the turbine and see how much the wood had worn. It had worn 1 inch in 82 years. When I got back up after being in a wetsuit I said to the lads, '*Alright you can lower it down it will see me out!*".

The new roller machines were designed to produce white flour. The corn was passed through grooved steel rollers turning at different speeds. They are known as break rolls and remove the outer skin or bran.

William Hipwell explained,

> "Instead of crushing the grain these machines scooped the kernel out and at the same time they cleaned the inside of the skin, the spiral grooves varied between millers according to how fine or coarse a separation you wanted."

Finally the kernel would be ground down to flour between smooth *reduction rollers*. These were generally, as at Holme Mills, approximately 20 inches long and 9 inches diameter.

The final stage of processing the flour is known as *dressing*. Most mills were equipped with some sort of dresser which was basically a sieve for removing impurities from the flour. The earliest form is the

Turbine power at **Stoke Mills, Sharnbrook.**

Beds. C.C. Photographic Unit

[1] **A Brief History of the River Ivel**, M C Evans, *Lock Gate*, Vol I.

A bank of roller mills at **Stoke Mills, Sharnbrook.**

Photo supplied by W. Hipwell

silk dresser. The remains of some are still to be found in Bedfordshire. This was a horizontal drum, usually three yards long covered with silk in a wooden case. The flour would be fed into the drum to be beaten through the silk so that only the finest flour was sent out of the mill.

A later improvement was the wire machine in which a series of rotating brushes swept the flour along a stationary cylinder lined with wire gauze. Power would be provided by belt drive from one of the main drive shafts.

In the few mills that continued to produce flour in the 20th century these silk and wire dressers were replaced by *plan sifters* otherwise known as *pendulum dressing machines* or, more affectionately as *drunken elephants*. Both Stoke Mills at Sharnbrook and Holme Mills at Biggleswade were equipped with them. *The Miller* of 4th May, 1896 contains the following description of the one at Holme Mills.

> "The Pendulum machine is composed of 11 sieves placed one above the other; in this particular mill each machine is arranged to take three independent feeds; the material enters at the top and passes alternately

over each sieve, the overtails leaving at the bottom of the machine. After passing through the silk the flour falls on the canvas fixed tightly over the silk of the underneath tray, and from there it falls to the bottom collecting tray, where the flour from all the sieves leaves at the same spout except when they are taking off Patent Flour, when the Patent goes down some of the spouts, and the remainder through the rest. It may be noted that when used on the Break Material these sieves are somewhat differently arranged, as, of course, there are more separations to make. In the Biggleswade Mill there are two sieves on the top for scalping, while the overtails pass to the succeeding break-roll. The next five sieves take out the flour, while the two following sieves grade the semolina, and the remaining sieves grade the middlings into two qualities ready for purifying. It will be seen, therefore, that the scalping, grading and dressing are successfully performed in one operation. It is claimed for the Pendulum that it differs from other machines of this type by requiring no device for cleaning the silks, such as brushes, grain or other material. The silks are hung loosely over the frames, so that the silks are kept in continual movement by the mere weight of the material which is being sifted; this prevents the flour from adhering to the sieves. The Pendulum is suspended from the centre of the machine, the object of this arrangement being to imitate by means of the peculiar shaking or agitating, and the eccentric movement of the crank, the motion of a sieve actuated by the human hand. There was no perceptible vibration caused by the five Pendulums; in fact if the visitor had shut his eyes, he would have found it impossible to detect any more motion than would have been perceptible from the use of a centrifugal installation of similar capacity."

However, the taste of the public for white bread meant that stone-ground flour was no longer popular. John Jordan explained,

"From about 1850 the bakers very misguidedly thought that if bread did not rise it was a slur on their ability to make bread. The presence of bran

Plan sifters at **Stoke Mills, Sharnbrook.**

Photo supplied by W. Hipwell

> impaired the baking quality so many millers introduced roller mills and threw out the bran for animal feed. But the big advantage of stone-ground flour is that it is full of enzymes."

So, to meet the mass urban demand for white flour, some mills went over to roller crushers. Now modern nutritional theory has shown the error of removing the bran rather than grinding wholemeal. Bran is still known as *offals* or leftovers, yet it contains the roughage vital to health. John Jordan continued,

> "Making white flour means giving the humans all the starch and rubbish and the animals all the vitamin and protein."

In the same article *The Miller* passed a comment in 1896 on Holme Mill,

> "Not the least significant among the signs of these times is the constant and never-ceasing progress of the Roller System among the country mills of this land. One by one the country millers are converting their mills to the most approved Roller System."

Roller mills are now operated so fast (about 600 rpm) that they generate heat and kill off all the enzymes in the grain. John Jordan said,

> "The big advantage of stone-ground flour is that it is cooler ground and is full of enzymes. We are trying to emulate stone-ground flour and have slowed our rollers down to 200 rpm."

Now wholemeal flour is coming back into fashion even though it means heavier loaves.

The Watermill at Biggleswade in about 1905, now converted into luxury flats. Beyond is Ivel House home of Henry Franklin and indicative of the prosperity that a miller could enjoy.

County Record Office

Chapter 3
Decline

J. Steele Elliott[1] wrote in 1931 about the county's windmills,

> "Upwards of 80 parishes in Bedfordshire have records of such mills in their history, whereas the present generation may not recall a tithe of them running. Biggleswade, Cranfield, Stevington and Keysoe remain, and they are working but intermittently. The others that are left are derelict and tempest shorn; in many instances they defy the elements by their wonderful construction of millwrights' workmanship and may serve a while for a few years longer as store-places or shelter for cattle."

Since those words were written, three out of the four windmills still running fifty years ago have gone: Stevington is now the sole survivor.

Over the country as a whole, the combination of improved inland transport and long-life flour took away the need for localised production of flour. The last operating windmill in the county was at Keysoe. The late John Filsell worked it from 1915 until the end of the Second World War. In 1975, at the age of 88, he remembered graphically why local milling disappeared,

> "We relied more or less on the little men, they would have an allotment, one part vegetables, a rood of wheat and a rood of barley for pigs. We used to do the grinding. My brother and me used to go round the villages and fetch it from them and take it back when we had ground it. The Government decided to subsidise farmers for wheat, oats and barley with tractors and harvesters. The subsidy was paid through the big millers. So the little men had to go into the towns to find jobs. And when they packed up we lost business."

Cranfield Windmill just before demolition in 1966.

Photo supplied by A. J. Wooding

Those water mills which could not be modernised usually switched from producing flour to animal foodstuffs. Flitwick Mill, for example, changed in 1914 to providing crushed barley for horse feed. Some of the more modern windmills, such as Henlow and Biggleswade were harnessed to steam and continued for a few years longer. But the local need was disappearing. John Goodman explained how it happened,

> "Every market gardener used to keep a dozen or so pigs. Now most farms have absolutely no livestock at all. There just is not the demand for animal feed."

By the time of the First World War many mills had ceased working altogether, no doubt having suffered neglect and undermaintenance

[1] *Bedfordshire Historical Record Society*, Vol xiv, 1931

for some time. The late Mr. Charles Scott of Flitwick recalled an event which must have taken place in the mid-1880s,

"There was a little mill down by Redborne School. And when I was three or four years old there was an old chap going down with the corn and, I living there, he would put me on the cart and take me for a ride. And I think I was taken down to this mill with the last lot of corn to be ground."

In contrast to the county's watermills, very little now remains of the windmills whose revolving sails were, until the turn of the century, almost as common a sight as the parish church. Four tower mills have been converted to houses and there are tangible remains of ten others.

By the 1940s and 1950s most windmills were greatly decayed. The next few years saw losses at an alarming rate. A typical tale of decay is that of the tower mill at Upper Dean. It ceased working in 1906. By 1931 when Steele Elliot's survey was published, it had lost two sails. By 1967 only two broken stocks were hanging on the pole end. The brakewheel was still in place but the ogee cap had gone; and the elements have since caused further damage to the remaining machinery.

Harnessing the wind has always been hazardous; there tends either to be too much of it or too little. Windmills were exceptionally vulnerable to storm damage and the mill left to decline in peace, like Upper Dean, is the exception. Most have met violent ends. They could, for instance, catch fire in high winds. If the wind caught the sails before they could be stopped, it could override the brake. This happened to the post mill at Yielden which was burned down in a gale in October 1877. Thc sails were forced into action and the friction from the brake set it on fire. Steele Elliot recounts,

"the miller being helpless it was burned out. The sails continued to revolve in the holocaust until the whole collapsed."

The remains of **Upper Dean Windmill.**
Photo D. Baker

Ben Anstee recalled that a storm had finished the working life of Cranfield Windmill,

"I was working there as a boy of 11 that day in 1935 when the mill got out of control in a storm. It has never been worked since."

Disaster nearly overtook the county's sole surviving mill at Stevington in 1919. Fred Harpin, the miller, vividly tells the alarming story,[2]

"It was one day in November when the guv'nor asked me to go up and grind a few bags of beans. A good wind was blowing so I clothed her up and got started. After a while the wind increased. I didn't like the way she was rocking about at all, so I called Arthur Coleman who was hedging nearby, and asked him to come up with me and there we stood rocking to and fro something awful. It was evident that something was wrong with her, and I said to Arthur: *'I'm going to stop her.'* I let down the brake lever and sat on it. By this time the sails were going round so fast that we both got scared stiff. Nothing happened, so Arthur came and sat on the lever with me, and there we were, going up and down like

[2] **A Windmill Saved**, Fred Woodbridge, *Bedfordshire Magazine*, Vol 4, 1954.

a seesaw, and I was afeared she might catch fire from the friction of the wooden brake on the headwheels as we couldn't slow her down at all. Arthur said, *'I'm not staying here any longer'* and took a header down the steps, but before he got to the ground there as a crash and the mill lurched forward, throwing him off the last half dozen steps. The sails had fouled the ground bringing the old mill up with such a jolt, so out I went double quick. *'That's the last time I'm going up there 'till this wind drops,'* I said. *'Nor me,'* said Arthur."

Stevington Mill in 1936 with sail cloths still in evidence.
Photo supplied by Harry Meyer

It was later discovered that a side girt (the main timber in the side of the body) had broken, causing the mill to fall forward. It was probably this accident that saved the mill from catching fire. The damage was repaired in the rebuilding in 1921 and the mill continued to work until it was honourably retired in 1938.

If a mill could be destroyed by a head wind, a wind from the rear could be even more damaging. The structure of the mill was designed to cope with a wind from the front. But if the sails were forced forward by a tailwind there were no bearings to hold the windshaft in place. The destruction of Keysoe Post Mill, the last working windmill in the county, is recorded in the Annual Report of 1948 of the Society for the Protection of Ancient Buildings,

"Since 1943 this mill has changed hands several times and again this year it was offered for sale. As the Society was making enquiries by telephone, news came through that it had been tail-winded and blown down. Mr. Rex Wailes visited and inspected the Mill and found the structure, although broken beyond repair, was sound enough and the accident would not have taken place had the mill been in the hands of a working miller who would have kept the mill turned to the wind or had the shutters removed from the spring sails. Keysoe was the only mill in England where the flour could be ground and dressed solely by wind power and it is particularly regrettable that this unique example should have been destroyed owing to lack of proper precautions."

Mr. Filsell, the last miller at Keysoe, had already noticed that all was not well,

"There was a very strong wind and it shot round and got behind it and toppled him over. Well, it was quite easy because it was going before. I was out for a walk towards the church and when I was coming back I looked across there and I thought, well that mill has moved. It's either moved the sails and the brake or it's toppling."

Keysoe Post Mill in 1928.
Photo Harry Meyer

If Keysoe Mill was destroyed by a human failure to turn the mill into the wind, Potton tower mill was destroyed by an unusual mechanical failure. Normally a tower mill, equipped wih a fantail, will turn itself into the wind automatically. Fred Woodbridge[3] has recorded what happened,

"At the end of December 1927 a north-easterly wind had turned the cap to face in that direction and then calmed off, leaving it in position. On January 6th, 1928, by one of those ugly whims of nature that every windmiller knows so well, a terrific south-westerly gale came up and the wind-vane, apparently frozen solid, would not move. Within a few hours the vane and sails were shattered almost beyond repair. It was not deemed economical to replace them as by that time the bulk of the

3 **Potton Windmill,** Fred Woodbridge, *Bedfordshire Magazine,* Vol 9, 1963.

milling trade was going to the great steam mills springing up about the county; so Potton windmill, once a pride of Bedfordshire, soon became derelict, another tombstone to a dead industry. It was finally demolished in 1955, leaving the tower half its original height."

In view of the almost unavoidable destruction of so many windmills through storm and tempest it is no less tragic to have to recount the loss of particularly fine examples at the hand of man. The demolition of the windmill at Biggleswade—latterly known as the *Steam Mill*—was particularly sad.[4] Acknowledged as the tallest and finest in the county and a Grade II listed buiding, it stood in Franklin's Yard off Hitchin Street. It had ceased working by wind in 1931. Its sails had become unsafe and were removed in the early 1950s. The remaining superstructure had deteriorated but the tower was still sound. In 1966 it was still used for grinding with the aid of electric motors but only had a capacity of 25-30 cwt. per hour, in contrast to 6 tons per hour by more modern machinery. The owners wanted the space it occupied for the more efficient use of the site and the brickwork for hardcore in its further development. They took the view that the mill was of no particular significance; others had been preserved so there was no useful point in keeping it. At 1966 prices £2,000 was required for its restoration. The public authorities were unable to contribute grants and in consequence could offer no more than token resistance to its demolition. The firm of Henry Franklin Ltd. was subsequenly taken over by Spillers and then by Dalgety. It was decided to close down the milling operations. So, ironically, the site only survived the sacrifice of the windmill by a few years. Now it has been redeveloped for housing in which the windmill could have made a most attractive feature.

Beaudesert tower mill was built in the early 19th century with four common sails, and winded by a chain wheel at the rear of boat-shaped cap. It is said to have had six patent sails in its final years before it was demolished at the end of the century.

The Luton Museum

Cranfield windmill, too, was demolished in 1966.[5] In this case the owners felt that its condition had become too dangerous. During the previous two years the roof had shifted considerably and the owners feared that local children would be tempted to explore it. But so soundly was the tower built that it resisted demolition to the last. Ben Anstee recalled,

"I think they had to charge it about 12 times before they could move it; the top was unsafe that's all."

In the post-war years the remaining flour mills became involved in a commercial battle between the large baking companies. William Hipwell recalled how Stoke Mills at Sharnbrook had been affected,

"After the war the small bakers gradually disappeared and we found that we were selling flour to the big millers and it was an extremely expensive job and they could call the tune."

In a bid to secure large parts of the market the *big four* — Ranks, Spillers, the Co-op and Garfield Weston — started buying up mills. John Jordan said,

"Within the last 20 years the big bakers bought up a lot of mills and shut a lot down."

[4] Sites and Monuments Records No. 908, Planning Department, Bedfordshire County Council

[5] *Woburn Reporter*, 20th October 1964

Stoke Mills, Sharnbrook in 1910.
Photo supplied by R. Izzard

Hipwell & Sons went in with Garfield Weston and flour was milled for a further 10 years until 1969, by which time the company had built a new roller mill at Tilbury. William Hipwell said,

> "He did not want any small mills left so he closed them down."

The arrangement in such cases was that the machinery should be scrapped and the mill *silenced* for ever.

The Ministry of Food had treated the largest plant bakery and the small local bakery on equal terms for subsidy purposes as part of government post-war policy at that time of controlling essential foodstuffs. The status of the local millers began to decline rapidly. We have already seen John Filsell's explanation for the destruction of local milling (Chapter I). The minutes of Hipwell & Sons tell the same story in more commercial terms. On 12th May, 1953, they record,

> "Many of the country millers' natural customers, the local small bakers, had been forced out of business in all parts of England during control by treating the largest plant bakery and the eight sack week family baker on equal terms in the matter of the rate of subsidy with the result that the *little* man had been unable to make a livelihood however hard he might work."

(These controls were not abolished until 8th September, 1956, when *National Flour* was ended.)

These fundamental changes towards more centralised baking started to hit the millers. On 12th March, 1954, the minutes read,

> ". . . the loss of some two-thirds of its previous customers amongst the country bakers who no longer existed due to the inequitable treatment by the Ministry as regard subsidy . . . every medium-sized miller, finding his mill on short time, started making flour for the biscuit manufacturers which resulted in the latter being able to buy flour at almost any price they liked."

One by one the great milling families of Bedfordshire gave up the unequal struggle. In 1964 William Simmons Ltd. of Leighton Buzzard was acquired by Hovis McDougall. Hipwell & Sons sold out to Imperial Tobacco in 1960 and Franklins of Biggleswade was acquired by Spillers Dalgety. John Goodman has kept Flitwick Mill going for animal feed but of the big millers only John Jordan remains. He said;

> "I ran this mill for years losing money. People often ask why I am the only flour miller left. Well, the only reason is because I like flour milling."

What are the prospects for restoring and preserving the remaining mills in the county? Some—notably Holme Mills, Astwick Mill,

Doolittle Mill, Totternhoe. An early photograph of this unusual combined wind and watermill taken before 1868 (the year when the sails were blown off)

Beds. C.C. Photographic Unit

Hyde Mill and Flitwick Mill—are still in working order and are not under any immediate threat. But the demand for maintenance of ageing structures is bound to increase as years go by. John Goodman doubted whether the upper floor of Flitwick Mill would stand the weight of the grain bins if they were ever filled to capacity as they used to be at harvest time. There is the danger that the small millers will not be able or willing to meet the overheads of keeping the mills going. Unfortunately there are few sources of public or charitable funds to which millers can turn.

One possible aim is the restoration of the remaining mills to full working order. This has, for instance, been achieved at Bromham Mill, acquired by the County Council who commissioned the firm of Gormley & Goodman to restore it. One of the waterwheels and a complete set of machinery driving two pairs of stones are now in working order: the wheel turns occasionally. The machinery in the watermill at Woburn Abbey appears to be capable of restoration relatively easily. Some students from the Architectural Association in 1978 reported that,

> "The principal gearing is generally in good condition. The failure of the floor timbers is the immediate obstacle to restoration. The upper stones are missing together with their enclosing vats, as are the wooden horses and hoppers required to feed the grain into the eye of the stones. Clearly these would have to be replaced. The condition of the leat outside the building and its controlling hatches is not known and it may be assumed that some repairs and renewals would be necessary. The flour dressing machinery which must once have been located on the first floor has all disappeared. Some replacement parts might be found. Complete restoration is a possibility. In view of the current interest in health food, stone-ground flour could find a ready market with visitors to the Abbey."

Doolittle Mill, Totternhoe, as it is today from south-west.

Beds. C.C. Photographic Unit

Of all the decaying mills in the county the most complete — and the one offering the best hope of restoration — is the combined wind and water Doolittle Mill at Totternhoe. The windmill part of the mill has not worked for over 100 years since the cap, sails and windshaft were blown away in 1868, so restoration would present considerable difficulties. However, most of the watermill is as it was left in 1924; only the waterwheel has badly deteriorated. Certain of the floors are rotten and one has partly collapsed. The late owner, Mr. Buckmaster-Linney, had been anxious to have the mill restored but had been unable to raise the necessary funds.

Useful work may also be undertaken with the more limited aim of protecting decaying machinery from further weathering. In such cases there is negligible chance of full restoration. But as there are so few mills left, such holding operations are worth attempting. Several such operations have been carried out in other counties but none in Bedfordshire. The only windmill meriting such work is the tower mill at Upper Dean, but even this is so far decayed that it may now be too late to salvage anything worthwhile. Campton watermill also is one that might have been saved but it is likely to collapse unless action is taken as a matter of urgency.

Where the machinery is hopelessly lost, the idea of finding a suitable use for the remaining shells may be justified. Several watermills have already been converted to houses. The conversion of Stoke Mills at Sharnbrook to a theatre is an imaginative and successful project. Conversion of the industrial mill at Biggleswade to flats has secured a future for another of the River Ivel's fine nineteenth-century mill buildings.

The machinery at Thurleigh tower mill has largely gone and the interior is said to be rotten and dangerous. Under the circumstances, a sympathetic house conversion might be considered to be an acceptable way of ensuring the retention of this unique Lincolnshire-style brick tower.

As always, the lack of resources prevents desirable conversion work on the surviving mills. Other counties have seen the formation of groups of enthusiasts willing to undertake voluntary labour to save the remaining stock of mills. This has not happened in Bedfordshire.

Although the experience of millers has been one of rapid retreat, John Jordan feels that the tide may be turning,

> "in the States they are going back to the village shop so that the bread goes straight from the bakery to the customer."

Indeed several local small-scale mills have reopened and serve the health food shops. Just beyond the Cambridgeshire border, at Over, Chris Wilson enjoys a healthy demand for wholemeal stone-ground flour from his restored tower mill. But it is difficult to see how this trend might bring regeneration to any of Bedfordshire's mills.

Interior of **Stoke Mills, Sharnbrook,** now converted to a theatre.

Beds. C.C. Photographic Unit

Chapter 4
The Miller at Work

The miller has not always been a popular figure. He has had a reputation for dishonesty and for exploiting his monopoly. Yet his skills in working a complex and often dangerous machine, at the mercy of wind or water, have perhaps had too little recognition. Everybody in the mediaeval village depended on him for their daily bread. He ranked third in line to the Lord of the Manor and the parish priest. His social position was paralleled by his economic position because the cost of having flour ground came third only to feudal dues and tithes. That the miller deserved his status there can be little doubt. Now he has passed into history. This chapter seeks to capture some flavour of this village worthy and his vanished way of life.

The machinery used by the mediaeval miller had gradually[1] evolved over centuries until the miller, standing before the hurst frame, was in full control of the grinding process. The meal, warm from the stones above him, was immediately to hand so that he could adjust the tentering gear. John Jordan explained the concept of the *miller's thumb*.

> "The miller used to check the tightness of the rollers using his thumb and forefinger checking the grist from the rollers. You can tell by the coarseness of the semolina or grist whether or not the roll needs to go tighter or looser."

The crook strings governing the flow of grain were also to hand as was the sluice gear which controlled the flow of water and hence the speed of the machinery. The windmiller in his more restricted surroundings likewise had everything to hand. It was possible for him to tie a rope to the lever, throw it out through a purpose-made hatch and control the sails remotely from the ground whilst he either canvassed or reefed them. This arrangement can be seen at Stevington.

The mediaeval mill was a remarkably sophisticated machine and it is little wonder that the operator was held in high esteem. John Reynold's says,

> "The post mill was one of the most daring and ingenious works of the mediaeval carpenter."

He worked with little more than the skill of his hands and was aided only by a block and tackle for lifting the massive windshafts, stocks and sails. He faced the problem of balancing the stones in structures strong enough to withstand the battering of the weather in a

[1] **Windmills and Watermills,** John Reynolds, published Hugh Evelyn, London 1970

The stone floor at **Stevington Mill** showing milling at its simplest. The massive brakewheel with its apple wood teeth fits tightly under the gambrel roof. Note the "clasp-arm" method of construction where parallel rather than radial spokes enclose the axle and are held tight by wedges. The brake can be seen tightly gripping the circumference. It was operated by the lever on the left on to which George Field and Frank Harpin had to throw their weight on the near-disastrous November afternoon in 1919 (see Chapter III). The iron wallower is in the middle of the picture meshed with the brakewheel. The "quant" on which it is mounted drives the upper stone which can be clearly seen where some staves from the vat are missing.

Sacks of grain would be hoisted up through the trap door on floor on the right by means of the chain just visible on the right. Grain was fed into the hopper from where it rolled down the sloping chute known as the slipper. Its angle could be adjusted to control the speed at which the grain was fed to the stones.

The rope on the windshaft could be attached to the stones so that they could be lifted by windpower.

Beds. C.C. Photographic Unit

position some thirty feet up in the air. This had to be done in such a way that the stones lay true and level no matter which way the buck was turned. The smooth running of the mill required extreme accuracy in the construction of the brake wheels, pit wheels, wallowers and gear trains. The millwright was pushing con temporary technology to the limit. It is therefore all the more sad that so little remains as evidence of his skills.

There are few people left who remember what it was like to work in a mill. Ben Anstee used to help his father in Cranfield windmill when he was a boy. He recalled how it operated,

> "The corn is delivered into the bottom of the mill. Then there is a pulley right through from the ground floor and you used to wind it up to the second floor. You used to pull it up the centre of the mill on a chain through the shutters in the floor. You used to shoot it into the big tapered bins. They empty into the centre of the two grinding stones. As it grinds the flour falls into shutes down to the first floor. And to open the sails to catch the wind you worked from outside; I couldn't do it because I wasn't big enough! There was a big drive wheel up at the top with a wooden brake and you had to put the brake on and off."

The work was hard and hazardous. Although almost all mills had sack hoists it was not always possible to use them, particularly during calm August harvest days when there was no wind or water power to spare. John Clover explained that a sack of flour weighed 280 lbs. Mr. Harpin could remember carrying sixteen-stone sacks up the steps at Stevington. When Robert Prew[2] was the miller at Sharnbrook one of his sons was killed falling from a ladder in the windmill while carrying up a sackful of beans.

[2] **The Mills of Sharnbrook**, Frank Hipwell, *The Sharnbrook Review*, 1949

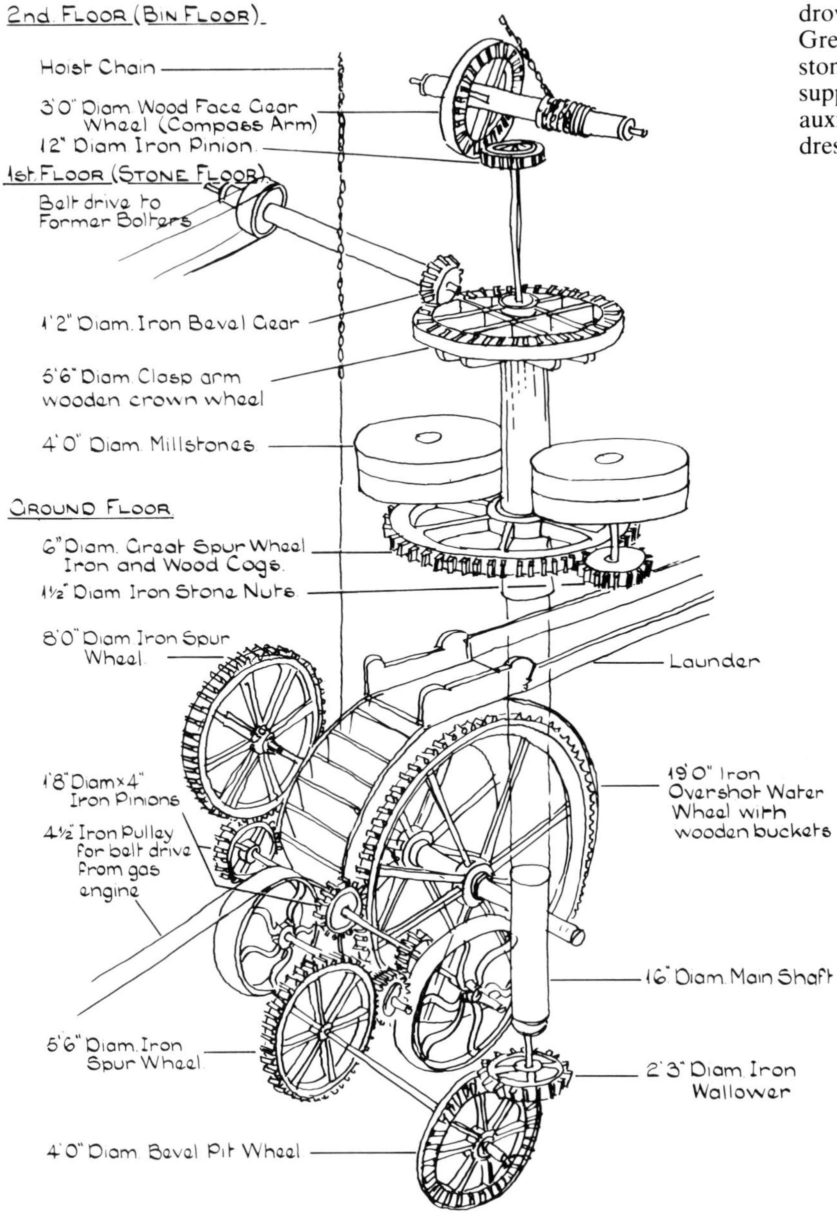

Details of the mechanics of the watermill at **Woburn Abbey.** Water flowed down the launder on to the overshot waterwheel which drove through a gear train consisting of pithead and spur wheels to the wallower at the bottom right of the picture. This drove the main vertical shaft. The Great spur wheel drove the stones. At the top the crownwheel supplied the power for the auxiliary equipment, the flour dressers and sack hoist.

The giant "on edge" stones for bone crushing at **South Mills, Blunham.**

Beds. C.C. Photographic Unit

Ben Anstee remembered working the mill was especially demanding at harvest time,

> "We've been there till midnight, I would come home from school and if the wind weren't blowing we had to start the steam mill up. We had a lot to do. That was the idea (of the steam mill) so that we could work in all weathers. I remember the old steam mill exploding and scalding my Dad and that's how I came to use the windmill. The steam mill used to get white hot and you had to walk past to start and stop it. My Dad was going by and it burst and scalded him."

Many were the injuries, often fatal, suffered by millers in days gone by. The miller certainly had to be quick-witted for he had to face personal risks of this nature every day of his life. He worked in cramped quarters, close to machinery that lacked the safeguards which are essential to any modern factory. The Coroners' Rolls[3] record the fate of a number of men who died in Bedfordshire's early mills. In 1269 for instance,

> "Thomas, son of Thomas Julian of Crawley, came to the Abbot of Woburn's mill called *Horsepol* in Husborne Crawley and took some grease to grease the lower wheel of the mill with his left hand and stood too close to it. The wheel held him and broke the whole of his left arm. The mill was stopped and he was rescued alive, but he died after noon on the same day."

At the heart of the mill were the millstones. Even people who know nothing about how a mill works have probably seen a millstone at some time, perhaps used decoratively as a door step. They are generally about four feet in diameter, blue/grey in colour, with a pattern of grooves on their running surfaces and with a hole in the middle. They are a subject of study almost on their own and there are many subtleties which may elude the casual observer.

Mr. Filsell recalled,

> "There were two different sets of stones, one for cattle offals and one for household flour."

Traditionally they are monolithic and are made from *Millstone grit* from the Peak District of Derbyshire. These were adequate for coarser grinding. But during the 19th century composite stones known as *French burrs* became more usual as they were capable of producing a much finer flour. They were more expensive: £36 bought a pair of French burrs in Victorian times compared with only £5 for a pair of Derbyshire peaks. Today a pair of French burrs would cost several hundred pounds. They consisted of many pieces of hard quartz, imported from France and made up by specialist firms concentrated in Mark Lane in London where the factors of machinery and millstones would receive their orders from the millers. Taylor's Mill, at Stotfold, contains an iron plate which recalls the name of one of these companies, *Millwright-Corcoram Manufacturing, Mark Lane, London.* Other such companies in Mark Lane specialised in other milling equipment such as bushel weighers and corn measures.

[3] *Bedfordshire Historical Record Society,* Vol XL. 1 p.40

Mr. Frank Hipwell[2] too has explained the distinction between these two types of stone, both of which can still be found in Bedfordshire's remaining mills.

> "It may be of interest to mention that the stones used were of two sorts. Those used for grinding wheat to flour were of French burr, imported from abroad in small blocks which were shaped and cemented together by specialists and subsequently bound with iron hoops to ensure that the runner should not fly apart when in action."

Disintegration was not unknown. John Clover explained,

> "There have been occasions when the hoops have broken and the stones more or less exploded, pieces would fly through the walls of the mill and if anyone was in the way it was too bad."

Coarse grain, such as barley, beans, etc., was dealt with on Derbyshire Peakstones quarried and shaped to form a one-piece stone. Furrows were cut from *eye* (centre) to *skirt* (circumference) of both upper and lower stones and formed the cutting edge. This was an easy task. The skilled part of stone dressing lies in *stitching* or *fluting* French stones and in getting the working surface exactly flat and even. The dress had to be renewed from time to time by the miller as the stones wore away. Far harder and more painstaking work was needed on the burrs than on the peaks. Frank Hipwell has written,

> "these furrows served to ventilate the surfaces which became very hot in action, and also gave direction to the meal in its travel from the centre of the stones to their perimeter. The lands between the furrows had also to be roughened with the millbill, and when dressing was complete and the runner lowered over the bedstone. The balance of the former required careful attention before it was put to work."

Both types of stones had what was known as a *dress* put in. Frank Hipwell was describing the treatment of Peak stones but French stones needed *stitching* with a millbill. The stitching consisted of fine parallel grooves along the *land* and served to grind the flour. Grains would be caught in the furrows and squeezed and rolled open, the *stitches* then scraped the flour off the bran and ground it.

Dressing the stone was a skilled and labour-intensive business. John Clover said,

> "You would have to dress the stones probably once a month, peripatetic craftsmen used to make a living from it. First they used to get a staff that was absolutely true, then they would make up some reddle (red lead) which would show up the high spots and then flatten them. Then they used to cut so many grooves to the inch."

The minute particles of steel from the cutting tool became embedded in their skin. This was the testimonial of their experience, and millers would ask them to *show their metal* before contracting them to work at his mill. The tool was known as a bill and its wooden handle was called a thrift. But these itinerant stone

The "luccam" of sack hoist at **Langford Mill** showing where sacks of corn would be hauled to the top of the mill through an upward-opening trap door.

Beds. C.C. Photographic Unit.

[2] **The Mills of Sharnbrook,** Frank Hipwell, *The Sharnbrook Rewiew,* 1949

dressers went out of business. Ben Anstee remembers helping his father to dress the stones in the 1930s,

> "I used to help my Dad sharpen the stones. We used to have a block and tackle, lift the top stone and lay it flat. Then we chipped them out with wedges with wooden handles. We used to sit on the stones and cut them out like herring bones. They were all that shape but some were coarser stones than others. Some we had to cut deeper than others depending on what you were grinding—beans, wheat offals or barley."

John Clover emphasised what tedious work it was,

> "You lay across the stone with a sack of bran under your elbow and then chip . . . chip. . . . chip. . . . If you rubbed your finger over them one way you felt the sharp edge . . the other way you didn't."

It was absolutely vital to the safety of the mill that the stones should be level and properly balanced. Each stone weighed about 15 cwt. and turned most efficiently at a speed of 125-150 revolutions per minute. The actual clearance required very fine adjustment. Maladjustment could mean flying sparks, with a serious risk of fire.

The stones were encased in wooden vats whose primary purpose was to prevent the flour flying about when it emerged around the circumference of the stones. Several good examples of complete vatted stones with hoppers and shoes can be found in the county, notably at Stevington and Totternhoe. Such an arrangement is described in the inventory to a lease on Barton Mill in 1907, *'Pair of 4ft French stones with hopper, ladder, damsel, brasses, cases and bill complete . . .'* The grain was fed into the centre, or eye, of the

John Clover demonstrates how to "stitch" or "flute" a millstone using a mill bill and thrift. The skilled part of stone dressing lies in getting the working surface exactly flat and making a series of fine parallel grooves which serve to grind the flour. Grains travel down the furrows and when caught, squeezed and rolled open before being ground in the "stitches". The sectional construction of a French "Burr" stone is clearly shown as are its iron bands.

Beds. C.C. Photographic Unit

runner stone from a hopper in which it was stored above the stones. The miller could control the rate of flow by adjusting the angle of the *shoe* or channel along which the grain flowed to the stones. The adjustment was carried out with a *crook* string which ran over a pulley down to the floor below where the miller would normally be. The bell warned when the supply of grain was running low and the danger of fire was greatest.

Fire was an ever-present risk for the miller, for flour dust can be highly inflammable. Clearly vigilance was the best protection against disaster. Mr. Goodman of Flitwick Mill explained,

> "If you were grinding on the stones and they ran out of corn and nobody noticed it then they will strike sparks and will set the wooden hooping around the stones on fire. I can remember it happening once. It just scorched the wood but there was somebody here and we caught it in time."

William Hipwell remembered the importance for fire prevention of good morale amongst the workers. They had won many bakery competitions and clearly were vitally concerned for the safety of Stoke Mills.

> "We had a marvellous lot of chaps. They had come down from generation to generation and if they saw somebody smoking they would push him outside the mill in no time at all."

A report in the *Bedford Mercury* on 17th February, 1844, illustrates just how disastrous a mill fire could be. In this case the victim was the mill that used to stand in Mill Street, Bedford.

> "Destructive fire — On Saturday morning last, about 2 o'clock, as we mentioned in our last edition of that day, the new wind and steam mill, the property of Mr. Attack, in St. Cuthbert's parish in this town, was discovered to be in flames, rapidly spreading all over the mill, which was principally erected of wood, and was situated in a yard leading into Newnham Street, St. Cuthbert's, and was used for grinding corn and sawing timber. The steam had been at work a quarter before 9 o'clock on the Friday night previous, and the mill was in it best working order. The flames soon spread all over the building and in the dead of night created quite a Vauxhall exhibition of fireworks and grandeur seldom to be seen. The cry of *fire* in the town brought hundreds of half-dressed citizens to witness the conflageration, which, had the wind been in an Eastern direction, would have extended along the south side of the street. Fortunately, the wind kept to the Northward and although large flakes of burning matter fell around the stacks in Alderman Higgin's field, at a short distance, they were preserved from destruction by tarpaulins spread over them, and men on the top, to watch and put out any falling fire. The first of the houses to the West, and adjoining the mill, occupied by W. Bettles, was ignited, and is materially damaged, but four engines having arrived, and a plentiful supply of water being obtained from Alderman Higgin's pond in the farmyard, the flames were soon overcome, and did not extend beyond the mill premises and Bettle's house adjoining, which was partially destroyed, but the mill is entirely consumed with the exception of its chimneys and some of the iron machinery, which is greatly injured. Damage to the amount of £800 has been done. The fire was first observed in the second floor of the mill, in the dressing room, and is supposed to have originated from

Interior of **Doolittle Mill, Totternhoe,** showing vat, horse, hopper and shoe.

Beds. C.C. Photographic Unit

A CHONDROMETER or bushel weigher. A vital part of the miller's equipment used for measuring the specific gravity of flour. It takes to pieces and fits into a small mahogany case for easy carriage.

Photo C. W. Howes

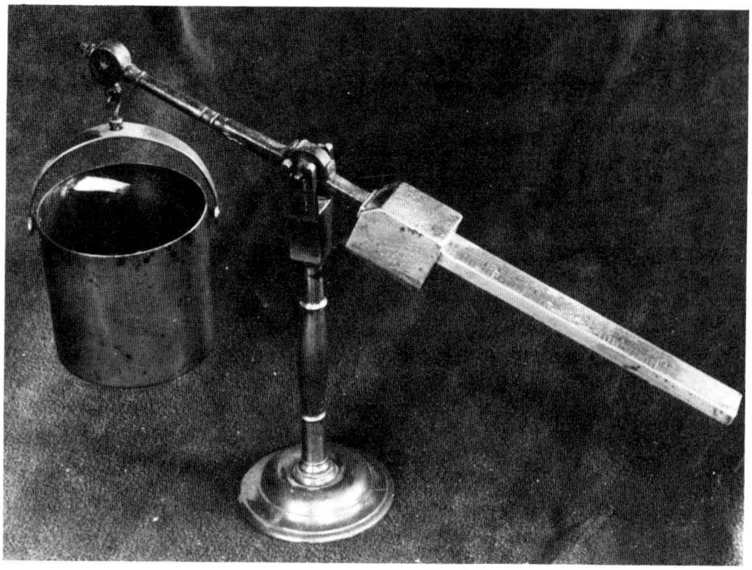

the friction of the machinery. The case of the owners is deserving of sympathy, for their all is destroyed. They are two brothers, millwrights, who from the use of iron millwork having rendered work at their trade extremely scarce, with a little property erected a windmill, and applied steam power to drive the machinery when the wind failed. They borrowed £200 on mortgage, which sum alone was insured; and this enabled them to complete the erection and works of the premises. The mill has never been idle half a day since it was erected two years ago, and was enabling the owners to retrieve their engagements but their all has been annihilated, and they have lost the means of a comfortable livelihood."

There were a number of Bedfordshire families in the milling business. Some met the challenge of changing circumstances with more success than others. The story of Henry Franklin himself is archetypal of Victorian success. When he died in 1932 an obituary[4] described him as

"a kindly and sympathetic employer who coupled with his directive capacity a full knowledge of his business. It was under his personal direction that the business had developed to the present magnitude and such an effort could only be accomplished by painstaking effort and the co-operation of the employees. *'He was the best master anyone could wish to have,'* said one of his oldest employees."

The obituary also told how he came to Biggleswade at the age of sixteen in 1867 and established a coal and corn merchant's business,

"As a corn miller Mr. Franklin was most successful with two mills in Biggleswade and one at Langford . . . Mr. Franklin still visited the Biggleswade, Hitchin, Bedford and London corn markets very regularly and few men were better known or more highly esteemed than he. By his industry and devotion to business he set up a fine example to the young men of today."

[4] *The Biggleswade Chronicle*, 24th April 1932

The Simmons family, too, were virtuously engaged for 120 years in flour milling in the south-west of the county. Richard Simmons started flour milling at Bellowes Mill in Eaton Bray in 1844 and thirty years later his son Frederick acquired Leighton Mills (known as Flemmons Mill) to be run by his son William after whom the company was named. Like the Hipwells and Jordans, he installed roller mills and a turbine at Leighton Mills in the 1880s. A disastrous fire in 1911 meant a complete rebuild with even more modern machinery. In the 1950s the company hit on the idea of turning its name upside down and the SNOWIS chain of retail shops was born. The bakery at Leighton Mills was opened. But now the end of milling was in sight. The company was taken over by Rank Hovis McDougall Ltd. in 1964 and within months the old mill was silenced. It was finally demolished in 1972/73.

But milling could also bring ruin, as the bankruptcy of George Powers in 1893 testifies. The family started milling on the Ivel in 1844 when Charles Powers[5] bought *"Water corn mills, called South Blunham and tackle etc. and Tolls on Boats using the water (Ivel Navigation Act) and dwelling house adjacent. Rent £125 p.a. £10 per every acre of sward ploughed up."* The various members of the family also acquired Holme Mill and the mills at Biggleswade, Langford and Stanford. For a while they prospered but in 1872 South Mills was completely destroyed by a fire which started in the area where oil cake and linseed oil was stored. The trustees of Charles Powers commissioned John Usher, the famous Bedford architect with other fine buildings to his credit, to rebuild the mill. It was to be financed by the proceeds of the insurance settlement which consisted of £7,500 for the stock in trade, £1,200 for machinery and £1,500 for the labour. But in August 1876 the mill was again destroyed by fire. Litigation then followed over the amount of the insurance settlement payable on the stock and goods destroyed. Eventually the mill was sold in 1898 and rebuilt for the manufacture of bonemeal.

It was virtually impossible to lift an 18 or 20 stone sack from the ground. This self lifting sack barrow enabled sacks to be cranked up to shoulder height.

Beds C. C. Photographic Unit

Meanwhile, in 1876 the Ivel Navigation was closed. With the coming of the railway in 1850 the river traffic had declined and with it the revenue from the threepence toll paid on each barge for passing the mills. The loss of revenue may have hastened the Powers' downfall.

Worse was to follow. New machinery had been installed in Tempsford Mill between 1881 and June 1883 but within six months this mill, too, caught fire. Hugh Powers, one of George's brothers, recorded his dismay in a letter to his solicitor dated 20th December, 1883:

> "Dear Mr Hooper, In reply to your letter I was thunderstruck when I received a letter from Walter to say that Tempsford Mill was destroyed and not fully insured. Since I have left Biggleswade I have had frequent conversations with my brother Fred as to the insurance of both mills and he always assured me that it was alright; so I have not thought it necessary to ask for a statement of the amount. . . ."

5 **The Powers Papers,** County Record Office

The late John Filsell demonstrates how to furl a cloth sail at **Keysoe Mill.**

Photos R. Wailes, 1936

The next year the Powers' partnership was dissolved. George Powers came to an arrangement with his creditors from 2 shillings to 10 shillings in the pound for debts of £15,000.[6] For some years he continued to trade with his brother Charles at Stanford Mills until Charles died in 1889. The end for George came on 1st December, 1893 in the London Court of Bankruptcy. He admitted to the Official Receiver that his recent transactions as a broker had been of a sporadic nature and his trading had resulted in a continuous loss. He had to agree that it was due to his want of judgement.

By all accounts the Powers were somewhat eccentric. The unfortunate Fred Powers, who later mishandled the insurance of Tempsford Mill, tried to close the footpath along the river by taking away a bridge that ran over a small stream at Biggleswade Mill. This created a stir in the town and a strong group of men collected outside the mill beating tin trays. They made an infernal row and noise and, in a frenzy, pushed the whole of the wall in front of the mill into the mill pit in an endeavour to stop the water that turned the mill wheel. Powers had to replace the bridge, and rebuild the wall.[5]

Rivalry between millers could give rise to the most capricious incidents. The supply of water for turning mill wheels was a frequent source of dispute between millers. The owner of one mill (who relied on the use of a turbine for providing domestic lighting) complained to me about the lack of co-operation of the next miller upstream who failed to indicate when he planned to hold up the water.

A miller needed a plentiful supply of water in the reach above his

[5] **The Powers Papers,** County Record Office

[6] *The Biggleswade Chronicle* 1893

mill to drive his wheel. But he also needed to get rid of the water below his mill. If water backed up below the waterwheel the power of the mill was considerably reduced. The minutes of Hipwell & Son record that the heavy floods of the winter of 1960/61 which raised the level of the Ouse below Stoke Mills prevented the turbine from functioning properly. And ice could bring the mill to a halt. John Clover remembered,

> "All night you would hear the wheel grinding away against the ice — and then it would stop."

At other times of year problems could arise at any time because one miller's headwaters were the next miller's tail race. This could be bad enough in ordinary circumstances but in the long-standing warfare between William Jordan and Henry Franklin *backwatering* became a critical weapon. John Jordan explained,

> "We were responsible for the banks of the river between here (Holme Mills) and Henry Franklin's mill at Langford. So what the old man did was to get all his men to raise the banks so that Henry Franklin had got a high tail water which would deprive him of his power. Then Henry Franklin would send blokes along to knock a hole in the bank!"

Finally, the miller was as subject to human frailty as anyone else. From the newly milled flour he deducted his portion. But he might take a little more and this is how he gained his reputation for dishonesty. He might, for instance, keep his thumb inside the measure and thus give short measure. For this reason he was said to have a golden thumb. Indeed as early as 1212 the men of Lecton (Leighton Buzzard) were complaining that more meal was being taken in tolls at the mill than should have been. And in 1574 when

A Straker-Squire steam lorry being loaded with 20-stone sacks of flour at **Kempston Mill** in the early years of the century. The driver is George Harris who won a gold watch for his feats in lifting such sacks. The owner of the mill, William Horn, is standing at the rear of the lorry.

Photo supplied by J. Clover

Robert Goodhall, owner of Duck Mill in Bedford, made his will he detailed a considerable list of valuable possessions.

> "Millers have always been credited with a golden thumb[7] and the clothes and furniture he left his family showed how comfortably he lived. He had a goblet of silver and gilt, a gilt salt cellar: he had tables and cottons and cupboards, forms and turned chairs, boarded bedsteads and feather beds, pewter dishes for daily use and silver rings . . ."

Frank Hipwell explained how the millers' bad reputation arose.

> "The miller's perquisite was such meal as might be left within the wooden casing of the stones, the interior of which would have been carefully swept out before the customer arrived. These cases were usually round in shape and fitted the stones fairly closely, but instances were not unknown (elsewhere) of square boxes which were more greedy in that extra meal settled in the corners."

Dishonestly can never have been so blatant as at Willington water mill[8] where the last miller's wife took a basin of flour for her pudding

[7] **Ouse History** — *The Lock Gate,* Vols 1 and 2 1962

[8] **Willington,** Joyce Godber, *Bedfordshire Magazine,* Vol 5, 1955

from every load of corn ground at the mill. In any event the unfortunate customer could return home with less material than he had brought.

On the other hand, one can spare a smile too for the unfortunate millers of Kempston. A custom[8] had become enshrined in law that the miller had to supply a cake at the court leet of the three main manors, no doubt for the refreshment of the jury. So incensed were its hungry members in 1466 at the miller's oversight in failing to supply the much-needed sustenance that he was fined 2d. In 1621 one of his successors was in equal odium. He had supplied a *great white cake* but the jurors rejected it on the grounds that it was too small!

The technology available to the miller could give rise to some bizarre incidents. The windmill's sails, for instance, offered the prospect of light-hearted, if foolhardy, feats. Steele Elliot recalled that at the Great Barford smock mill a man was lashed head downwards on to the bottom sail, and then allowed to go round one revolution. Steele Elliot says,

> "We can but hope that he felt adequately recompensed by the view obtained from above."

The late Mr. Filsell recalled a similar incident at Keysoe,

> "Charlie Woodford went over on the sail."

Why did he do it?

> "Just for a bit of foolery."

Shuttered sails at **Keysoe Mill.**
Photo R. Wailes

John Filsell at **Keysoe Mill** demonstrates how to turn a post mill to face the wind.

Photos R. Wailes 1936

No doubt he would have had second thoughts had he known the speed of the sails. In Mr. Filsell's words they were,

"Faster than any train on the Sharnbrook line."

The feud on the Ivel between Henry Franklin and William Jordan also had its humorous side. To conclude these millers' tales, let us imagine the reaction of poor Henry Franklin aroused from his well-earned slumbers by a wall of water set loose down the flooded Ivel by his neighbour. John Jordan recalled the details,

> "You would think they were very friendly but they weren't. If Henry Franklin sold flour for 5/2d a sack William Jordan would do it for 5/1d and then Franklin would undercut him again. And it got worse than that. When there was a lot of rain the old man would get his butler, a fellow called Pedder, and say, *'Go down the river and tell me when Mr. Franklin puts his light out.'* Pedder would come back at 11.00 at night and say, *'Henry Franklin has got his light out.'* So my grandfather would say, *'Right, Pedder, open the sluices; we'll flood old Henry Franklin out!'* Henry Franklin bought Langford Mill (the next one up the river) they say, solely for the purpose of flooding William Jordan out! They really had some fun in those days!"

Keysoe Mill

Chapter 5
Windmills

There are records of at least one hundred and forty windmills in Bedfordshire at one time or another but only Stevington Mill has survived in a complete state. Fortunately some eighteen others still existed well into this century.

Stevington represents the earliest type, the *post mill*. These were built on a massive cross of four oak beams which supported a large central post. The body of the mill, often known as the buck, could be rotated around this post, enabling the miller to turn the whole mill so that its sails could catch the wind from whatever direction it might come. Other examples of this early type stood at Bolnhurst, Lower Dean, Keysoe and Riseley.

The second type of windmill was the *tower mill*. This was an improvement over the post mill in that the main body was rigid, only the cap and sails turned to face the wind. They had brick or stone cylindrical towers and often occupied sites where post mills formerly had stood. They had fantails, a device dating from about 1745 which automatically turned the sails into the wind.

They also had *patent* sails, invented by William Cubitt in the nineteenth century, a feature that made the miller's life both simpler and safer. They replaced the *common* sails of the type that can still be seen at Stevington which had to be covered with canvas

The main post at **Stevington Mill** which supports the rotating body (or buck) of the mill; taken from inside the roundhouse.

Beds. C.C. Photographic Unit

Now converted to a house, **Stanbridge Mill** as it was in 1934.

Royal Commission on Historical Monuments, (England). HES Simmons Collection

and *reefed* by hand to suit the varying winds. This was a laborious process which could not be carried out whilst the mill was running. *Patent* sails, in contrast, consisted of rows of spring-mounted shutters which opened automatically when the wind became dangerously strong. They could also be controlled by a chain hanging out of the back of the windshaft low enough for the miller to reach it from the ground. With the exception of Sharnbrook Mill the tower mills of the county had fantails, and apart from Henlow, so did the smock mills.

The third and final evolution of the windmill was known as a *smock mill* because early examples, clad in canvas, were said to resemble a shepherd in his smock. They were octagonal timber towers built on brick bases and were said to be of Flemish design. They were essentially a cheaper version of the tower mill but usually enjoyed the sophistication of a fantail and automatic sails. There were at least five smock mills in the county, at Flitwick, Great Barford, Henlow, Riseley and Biscot.

Windmills had to be kept head into the wind to work most efficiently. In the old post mills, like Stevington and Keysoe, the whole body of the mill was turned by hand. John Filsell remembered,

> "It had a tailpole, you see, and we had to push it round by that. You got your weight behind it until it was facing into the wind. The thing was to start soon enough and not wait until the wind got too much so that you had to push against it."

The tower mills at Duloe and Stanbridge have now been converted to dwellings. The truncated remains of Hulcot Windmill (NGR SP 942382) have now been annexed to a modern dwelling. Most of the brick tower had been demolished in about 1880 when the machinery was taken to Spaldwick Mill in Huntingdonshire.

A smock mill once stood on the west side of Woburn on the Duke of Bedford's estates (NGR SP 941329). Like the *pug* mill it appears to have been built more as a technical experiment than for any normal purpose and was equally short-lived. As early as 1805 the timber work was taken down and the two-storey octagonal base was converted into a cottage which still stands.

Sharnbrook Mill is now an observation tower. And there are substantial ruinous remains of tower mills at Upper Dean, Dunstable, Houghton Conquest, Houghton Regis, Potton, Thurleigh and Totternhoe. The tower mills at Biggleswade and Cranfield have, as recounted in Chapter III, disappeared without trace.

Most of the Bedfordshire post mills were simple structures lacking the cross-bracing characteristic of those in the open, windswept expanses of East Anglia. The framing of the body followed the Midland practice which is not as well braced. Stevington Mill is the exception and perhaps, because of its superior construction, the only survivor. The body of the buck has a much better frame than the general run of post mills. The carpenter in charge of the rebuilding of the mill in 1921, the late P. W. Keech, studied French post mill construction whilst he was in the army during the First World War and applied the principles that he had learned to good effect at Stevington.

It is not usually possible to indicate with any great precision when most mills were built. Often the best estimate must be based on whether a mill appears on Jeffery's map of 1765 or Bryant's of 1826.

The oldest of the score of windmills to survive until recent times is **Lower Dean Post Mill** (NGR TL 041698). The absence of diagonal bracing in the construction of the buck is thought to have contributed to its collapse in 1959. Now only the base remains. The mill was built in 1762 and was substantially rebuilt in 1834 as a most unusual post mill called a composite mill. The buck ran on a circular cogged track around the top of the roundhouse. The fantail was geared to this track. The buck could be turned either by tailpole or fantail. Latterly the fantail failed owing to the body settling heavily on its rollers. Finally, the only way to pull the tailpole round was with the aid of a donkey.

An early photograph (circa 1905) of the unusual mill at **Lower Dean** whilst still in working order.

Photo supplied by Peter Dolman

Keysoe Post Mill (NGR TL 077627) was built in about 1800. It was a typical early example with a roundhouse. This was the last working windmill in Bedfordshire. By 1935 it was the only one where flour was still ground and dressed solely by windpower. Its roof was of the primitive straight pitched type and its body flared out at the bottom to form the roundhouse roof. Mr. Filsell explained why the roundhouse had been added,

> "The brickwork was only put round to keep corn and to shelter it from the rain. It was open at one time and the mill itself stood on a big post which was a tree on its own really."

The whole structure was tarred. The mill was turned to face the wind by pushing on a tailpole using a yoke after the steps had been raised by pulling a lever known as a *talthur*.

To tell the miller if the mill was pointing into the wind there were two wind indicators on either side of it and a fox windvane on the roof. A typical mishap occurred in 1935 and was described in the Year Book of the Society for the Protection of Ancient Buildings 1935.

Keysoe Mill showing both furled sail cloths and shuttered sails.

Royal Commission on Historical Monuments, (England). HES Simmons Collection

> "Last January a sail stock broke while this mill was at work. One sail was thrown clear of the mill damaging all the other sails in the process. The condition of the mill is such that at present it can no longer be worked. The two brothers who own the mill desire to make and fix the new sails themselves with local help and a generous offer has been received to supply the necessary timbers at cost price if this is done. Approximately £50 is required to complete essential repairs. £22 10s. has been subscribed at present of which £10 is the gift of the Society."

Its loss was especially sad as similar post mills in the area at Bolnhurst (NGR TL 094602) and Colmworth (NGR TL 112574) had already been demolished.

Stevington Post Mill (NGR SP 992528) was built in about 1770. Its body was totally rebuilt in 1921 and is not a replica of the original. Unfortunately no photograph has yet come to light showing how the mill looked before 1921 but perhaps it resembled those at Keysoe and Riseley. All the buck is new apart from the crowntree, the stones were replaced, the brakewheel is second-hand and the *tentering* gear is new; all that remains of the original mill are the post and trestle, crowntree and windshaft.

This mill stands on high ground at the eastern end of Stevington village. The roundhouse is built of stone and is roofed with tiles. The body of the mill is two storeys high and clad with black weatherboarding. It was turned into the eye of the wind by the miller pushing on the tailpole after he had levered the access ladder off the ground. It has a gambrel roof of corrugated iron.

In the early years of its life it had no roundhouse. This was added at some time during the late-19th century, and the stone piers were raised to their present height at the same time. In the stone paving around the mill Peter Dolman has identified fragments of burr stones and also some pieces too dark to be from the Peak District, so it could be German *Cullen* stone. This type of millstone was widely used in East Anglia and was highly regarded by millers for general

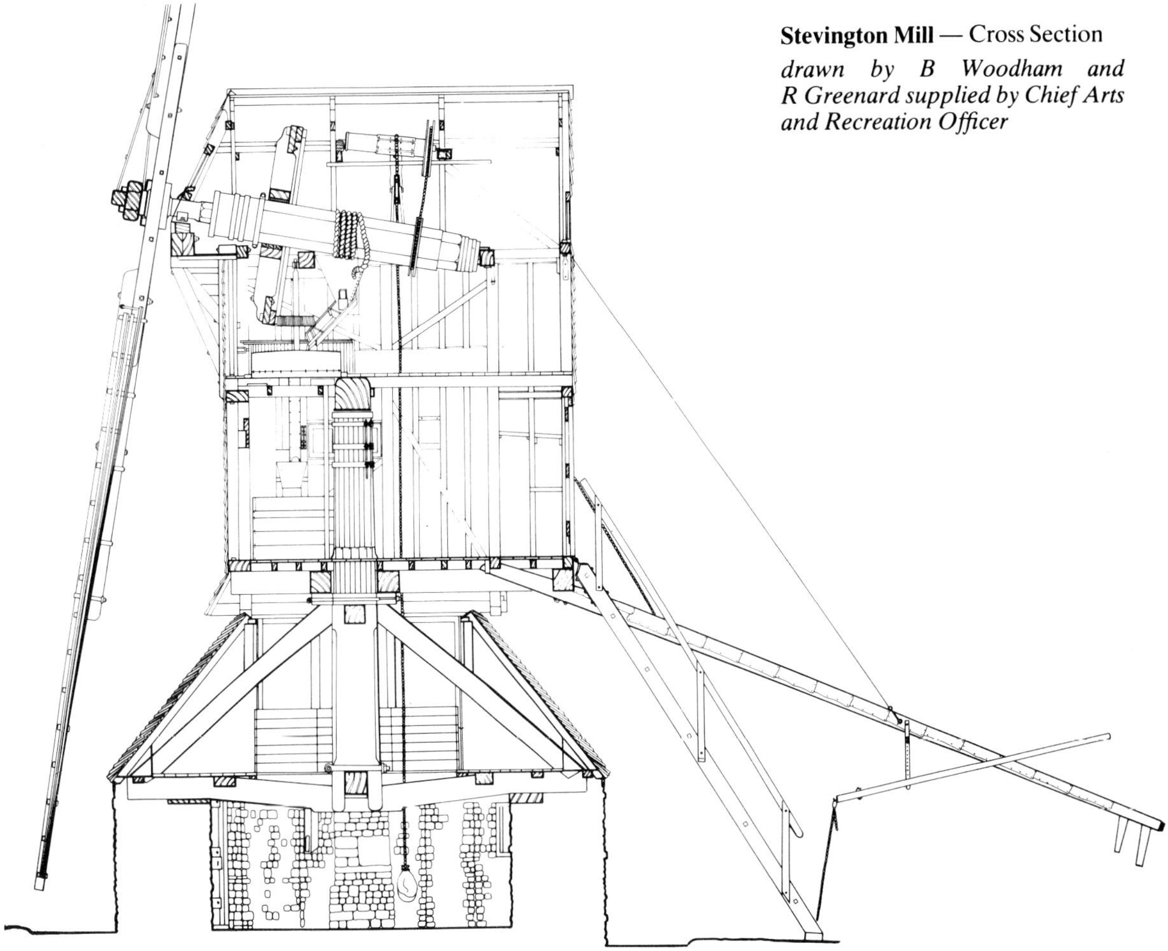

Stevington Mill — Cross Section
drawn by B Woodham and R Greenard supplied by Chief Arts and Recreation Officer

grinding. The mill was originally tarred, and the trestle is still coated in tar, although it has been enclosed for many years now.

The rebuild of 1921 was necessary after part of the frame had collapsed in 1919 while the mill was working (as detailed in Chapter III). It was decided to repair the body, retaining the trestle, crowntree and windshaft. A derrick was built over the mill, formed from four tall trees grown on the estate of Lord Ampthill at Oakley. The whole frame was then lifted about 18 inches and was carefully taken to pieces. The new body is carried by two pine beams of huge proportions, being 10 ft. long, 16 inches deep and 12 inches wide. In addition to the repair of the frame, a number of smaller jobs were carried out at the same time. A pair of peak stones were obtained from Milton Ernest watermill and were fitted, together with a new tentering device, by a millwright from the firm of Thomas Course & Sons of Bedford. A new oak pintle was fitted to the top of the post, as the old one had worn unevenly. The crosstrees were also raised at one side to bring the post vertical. The old millstones were placed at the back of the mill to counterbalance the sails. One has since been transferred to the roundhouse floor as an exhibit. There was a near disaster during the course of the rebuilding. The windshaft was being lowered into position when the tackle broke. Luckily the shaft dropped straight on to its bearings without damage to the rest of the structure.

From 1800 onwards smock and tower mills became increasingly common. A small tower mill, built at **Stanbridge** in 1800 (NGR 962254), was the earliest. It is 23 ft. 6 ins. over the base, with 24 ins. thick walls. To the top of its ogee cap it was 42 ft. high. It had a fantail and two patent sails of 60 ft. span and two common sails of about 70 ft. span. The patent sails were double-sided and were 6 ft. in width. There were four floors, the stones (two pairs) being on the first floor. It was gutted and converted into a house in the early sixties.

Duloe Tower Mill (NGR 170602) is a large and solidly built tower mill, of a rather quaint and unusual design. The tower is of brick, 24 ft. 6 ins. across at the bottom, 20 ft. across at the top or curb, with a wall thickness of 28 inches tapering to 18 inches at the curb. This was of wood, forming a track for the cap to turn on. Its almost cylindrical tower and conical cap are rare. Whilst there are some mills with

The tower mill at **Duloe Hill** in 1936.

Photo H. Meyer

Houghton Conquest tower mill in 1933. The tower still stands but is in a ruinous condition.

Royal Commission on Historical Monuments, (England). HES Simmons Collection

similar towers in Somerset the conical cap was unique in Bedfordshire. It has since been replaced by a more modern domed roof.

The mill was built in 1815 to replace an old post mill, which had been built in the 1700s and was burnt down in 1814. It had been in the Arnold family since at least 1785 and the present mill was built for William Arnold. His family had it until 1864, when it was sold to James Topham who sold it in 1873 to James Green. It was last worked in 1916 by Albert Green. The shutters were then removed to stop the sails turning. The fantail soon broke up and the mill remained with no shutters and fan until the late 1940s; then the sails and fan stage together with almost all the machinery were removed and the tower was made into a house. The only part of the machinery remaining is the huge oak windshaft. The 10 ft. diameter brakewheel is wedged on to the squared roof and the cast-iron boxes which carried the sails are fixed to the front of the shaft. The boxes (or poll-end) can be seen, sticking out of the cap.

Houghton Conquest Tower Mill (NGR TL 047432) replaced an 18th-century post mill, probably in the mid-1830s. It is a small tower mill 35 ft. high and built of brick. The shape of the walls is similar to that at Dunstable, being perpendicular up to first-floor level, and tapering above. The cap had a ball finial on the top. Due to bad design, the tower has cracked causing the window arches to fall out—the tower is effectively in two halves. The mill originally had common sails and was winded by hand, with wheel and chain gear. In October 1877 a heavy gale blew the complete cap away and smashed it. The mill was refitted with a new iron windshaft, patent sails and a fantail. Production ceased in 1910 and the sails were taken to Cranfield Mill. The mill is now in a ruinous state and the roof was removed in 1975 following a collapse. The remaining machinery has been put in store for future use in restoring other mills in East Anglia.

Riseley Post Mill (NGR TL 046625) may have been the last post mill to be built in the area. It was originally an open trestle mill, having a roundhouse added in the latter part of the 19th century. It was fairly large and not as primitive as some of the post mills in the area.

The remains of **Potton Post Mill** (NGR TL 213498) which was built in the 1840s, can still be seen today. It was advertised in the *Bedfordshire Times* in January 1853 as a,

> "First Rate Tower Windmill — All that remarkably well-built and nearly new freehold Tower windmill, with two pair of French stones, Six sails and fantail, in full trade, etc."

The six single-sided patent sails were blown off in 1879 and were replaced by four double-sided patent sails. The shaft could be driven by steam especially in calm weather. The mill, which was five floors high, was connected to a granary by an iron catwalk at first-floor level. The stones were two pairs of Burr, one of Peak stone and one of Emery composition. Other machines in the mill were an Oat Crusher on the third floor, and a wire machine and scourer on the second floor.

Riseley Post Mill in 1936.
Photo supplied by H. Meyer

Cranfield Mill (NGR SP 960429) was a tower mill of red brick with a cemented finish. It stood about 55 ft. to the acorn finial on top of its broad *ogee* cap. Around the base of the cap was a scalloped *petticoat* which stopped rain from getting in. The millstones were two pairs of *Peaks* and one of *French Burrs*. The four double-shuttered patent sails of about 65 ft. span came from Houghton Conquest Mill in 1910. The original sails were much larger, with a span of over 70 ft.

Dunstable Tower Mill (NGR TL 014218) is a large tower mill of Caddington *grey* brick, 28 ft. diameter at the base and with 3 ft. 6 ins. walls at the base and 2 ft. walls at the curb. The tower is unusual in that the first two floors are built vertically, the taper starting at second-floor level. This invites comparison with Houghton Conquest Mill, of similar shape but smaller. The mill is 58 ft. high to the top of the cap, which is domed and covered with copper. There was formerly a ball and stalk finial, and a scalloped wooden petticoat on the cap. The mill had four double shuttered patent sails and was *winded* by an eight-bladed fantail. Three pairs of stones were fitted, these being powered by gas engine in the 1920s. The mill was surrounded by many outbuildings, including a steam mill. In 1942 it was converted for use as a Sea Scout Headquarters, and repairs were carried out by them in 1956, mainly to the roof.

Shefford Tower Mill (NGR TL 149395) is a small mill, 40 ft. high, of red brick. It ceased work in the 1880s and became derelict. During the First World War, the copper was removed from the roof, causing rot to accelerate. By 1929, the roof had completely gone, and only parts of two sails remained. About that time the cap was taken off and, after remaining empty for a few years, a roof was put on and since then the tower has been used as a fruit store.

The smock mill at **Great Barford** (NGR TL 123532) was built for Edward Trustrum in about 1835. The millwrights were probably Thomas and William Course of Bedford, who also built a smock mill at Flitwick at about this time. It stood about 25 ft. high to the top of its conical cap, which had a fantail at the back for winding the mill. There were four floors, including the cap, and there were two pairs of stones, one of Peaks and one of Burrs, on the first floor. At a later date, a flour dresser was fitted and the mill was adapted for steam drive. The power was provided by two spring and two common sails.

In the *Bedford Mercury* of November 4th, 1837, the mill was offered for sale by auction and was described as being,

> "erected but a short period, very substantially, upon the best principle and is doing an excellent business."

It was again offered for sale on December 12th, 1897, together with a farm, house, granary, etc. It made £430 — not a very high sum for a mill. It was fully described in the sale notice at that time,

> ". . . a brick and timber three storey tower windmill working two pairs of stones dressing machine, all the going gear, pair of cloth and pair of spring sails fantail, pulley, pinion for steam power, sack tackle etc."

It ceased operation about 1900, and was sold in 1905 to W. R. Green. He had it dismantled in 1908, leaving the octagonal base, which was 15 ft. across, as a shed. This too was cleared away in 1925.

Flitwick Smock Mill (NGR TL 027355) is known to be the work of Thomas Course of Bedford in about 1850. The *Bedfordshire Times* for 28th February, 1852 contains the following as part of a sale notice:

> "All that new corn windmill situate near Dennel End aforesaid, fitted with one pair French and one pair Peake 4 ft stones and flour machine . . ."

A portable steam engine was kept in a corrugated iron shed near the mill and this was used to power the machinery during calm spells. In October 1881 it was advertised as *a windmill to let . . . in good working order.* Milling ceased shortly before the mill was burnt down on November 9th, 1903. The base was demolished and the bricks used to build stables which still stand. Visible in their walls are black bricks, which had been used to face the base of the wall and were therefore tarred.

Potton Tower Mill in 1926.
Photo H. Meyer

Dunstable Mill in its heyday in about 1900. Its four patent sails drove three pairs of stones. The chimney indicates that the mill could be driven by steam when the wind failed.

Photo Bedford Historical Record Society

Flitwick Smock Mill at the turn of the century

Photo supplied by Mr. Gregory, Windmill Road, Flitwick

Only the brick base of **Henlow Smock Mill** now remains (NGR TL 175389). Photographs suggest that it too was built by the Course brothers. It was used for grinding by wind power until 1916 when flour making for bread ceased. The sails were removed in 1925 after being damaged in a storm the previous year. It was worked by steam power from a portable traction engine until 1935 when the mill was demolished. There was no fantail — the cap being winded by a chain wheel and gear train protected by a built-out penthouse. This suggests that this mill was rather older than the Flitwick and Barford ones.

A brief mention should also be made of **Biscot Smock Mill,** Luton (NGR TL 197231) which replaced a post mill burnt down after lightning struck in 1841. A number of photographs record its disintegration after 1925 when the sails and fantail were removed. It was pulled down in 1938 when the site was needed for building a public house called appropriately *Biscot Mill.*

Upper Dean Mill (NGR TL 041082) is a small red brick mill, three storeys high, and formerly 46 ft. high to the top of the acorn finial on the cap. After stopping work it gradually became derelict. Small

Henlow Smock Mill in 1926. After it had lost its sails it was powered by steam. Now only the brick base remains.

Photo H. Meyer

Biscot Smock Mill in 1935 in its final years.

Photo H. Meyer

Sharnbrook Tower Mill in 1937 showing the unusual tailpole.

Photo H. Meyer

items like governors and hoppers, chutes, etc. were taken away. The metal on the cap was removed at some time and from then on the deterioration was rapid. The roof has now gone, and the cap frame is breaking up. The fanstage fell off c. 1960. The top floor has gone, as have all doors and windows. Two sails remain, the stock being in good condition still. The frames are in poor condition and all the shutters have gone. The brake has broken so the sails turn in high winds. which does not do the brakewheel much good.

Biggleswade Tower Mill (NGR TL 188441) the finest mill in the county, was erected in 1860 for Ezra Garner. It was 70 ft. high to the top of its cap and was of the Cambridgeshire type of tower mill, with an outward-flared *petticoat* to its ogee cap. The stones were on the third storey. The cap had a gallery around it and when the mill needed tarring, a cradle was suspended from it for the purpose.

Sharnbrook Tower Mill (NGR SP 998593) is a small tower mill, built of local oolitic limestone. There were three floors above ground. There were two pairs of 4 ft. stones on the first floor. Frank Hipwell[1] explained and unusual feature;

> "One curious feature of the building, due possibly to the desire to save expense in erection, was the method adopted for turning the sails into the wind. It was out of date even at that time, and is the only tower windmill in the County which has a tailpole attached to the revolving cap carrying the sails. These had therefore to be man-handled from the ground to correct their position. The usual method consisted of a fantail mechanism fixed on the opposite side of the cap, which automatically moved the sails into position when the wind veered, on the principle of a weather-vane."

[1] **The Mills of Sharnbrook,** Frank Hipwell, *The Sharnbrook Reviews 1949.*

Thurleigh Tower Mill (NGR TL 047582) is a fine tower mill of the Lincolnshire type. The height to the top of the cap was 60 ft. The cap was ogee or *onion* shaped. The tower is of brick, cemented over. There are five floors above ground level. On the second floor there are two pairs of stones, one of peaks and one of burrs. There is also the remains of a wire machine. The brakewheel, still visible, is about 9 ft. across and is completely of wood. All other gearing was of iron, with iron shafts. The mill was also fitted up for steam drive. One sail was blown off in 1917, and the other three, together with all the ironwork, were removed for scrap in 1940 *to help the war effort.* Removal of the upright shaft and gearing was easy, as it was all bolted together in sections. To remove the windshaft, the brake-wheel was loosened, a cable was attached to the cross, and the whole lot was pulled out by a tractor. The mill now presents a forlorn appearance, for the roof collapsed in 1969. The fanstage still remains in place but the mill is getting into a dangerous state. At the time of writing permission has been granted for it to be converted into a house.

Lastly, there are the meagre remains of the mill at **Houghton Regis** (NGR TL 01238). This was a typical *modern* tower mill which replaced a smock mill at an unknown date. It had four automatic double-sided patent sails, and a six-bladed fantail. It ceased working during the First World War and was dismantled in the 1920s. Only about 5 ft. of the brick tower were retained as a support for a water tank.

The mill ceased work about 1890. The sails were removed in 1920, and after remaining derelict for many years, its contents were stripped out in 1967 by the late Mr C. A. E. C. Howard, a local builder. He installed a staircase in the shell, and built an observatory, and installed an ancient clock in the top of the mill.

Thurleigh Tower Mill in 1936.

Photo H. Meyer

Overleaf

Franklins tower mill at Biggleswade latterly known as the 'Steam Mill'. Fantail and sail shutters are missing suggesting that the picture was taken during its 'steam' era probably during the 1930s.

Times Newspapers Ltd

Chapter 6
Watermills

The watermill has been evolving for nearly two thousand years. Though we know little about those that predate the eighteenth century, a wide variety survives in the county. For the twentieth-century citizen, accustomed to standardisation, the diversity of watermills can be most refreshing.

The 18th-century examples were usually constructed by local builders using local materials. They have a strong affinity with the agricultural architecture of the period and they are usually built of local red stock bricks and are clad with weatherboarding. In strong contrast the nineteenth-century mills, notably those in the Ivel valley, are built in the *industrial* tradition. They are large, factory-like, mostly of four storeys, and are built of white or yellow gault brickwork with roofs of Welsh slate.

Watermills are classified by the way that the water strikes the wheel. The most efficient kind is the overshot wheel where water is fed along an artificial channel, known as a *launder,* on to the top of the wheel. Excellent examples can be found in Bedfordshire at Barton, Flitwick, Woburn and the broken remains of one are still in evidence at Doolittle Mill in Totternhoe. Other examples existed at Maulden and Clophill. This system requires a rapid rate of fall of the stream to achieve at least 10 ft. difference in level of the water between the head of water above the mill and that below; thus it is normally only appropriate to hilly areas. But, by dint of damming up Bedfordshire's more sluggish streams to form substantial millponds, an appreciable head of water could be achieved so that overshot wheels could be used. In view of the very modest amount of water in some brooks the ability to use these most efficient wheels was a great asset.

"Overshot" waterwheel at **Barton Mill.**

Beds. C.C. Photographic Unit

The other common type of wheel found in Bedfordshire is the breast-shot wheel where a sluice directs the water into the buckets just below the midpoint of the wheel. This type was the subject of considerable scientific development during the 19th century. Initially, more efficient scoop buckets were fixed to existing wooden wheels. Later the wheels were cast entirely in iron. The newly restored wheel at Bromham Mill is a particularly fine example.

If the mill itself could be built over the river the wheels would be in the building itself as, again, at Bromham. Alternatively, if the mill building had to be sited beside the river, the wheel would be outside the building and mounted on an axle which would extend through the end wall of the building to drive the machinery, as at Flitwick.

Of Bedfordshire's rivers only the Great Ouse can be regarded in any sense as mature, having already flowed some 30 miles from its source near Brackley before entering the county at Turvey. Thus, in contrast to mills on other rivers, the Great Ouse mills were better assured of a constant and adequate supply of water. The Ivel, for instance, rises only a short distance over the Hertfordshire border whilst the Lea, Ousel and Flit have their sources in Bedfordshire. They are therefore little more than youthful streams of uncertain flow. Millers often had to use their water with the utmost economy to keep their wheels turning.

"Breast-shot" waterwheels at **South Mills, Blunham.**

Beds. C.C. Photographic Unit

The Ouzel Mills

The Well Head stream, the headwaters of the Ouzel, rises under the scarp slope of the Chilterns. Within 400 yards of the source lies **Doolittle Mill,** one of the most interesting mills in Bedfordshire (NGR SP 990202). It is a rare example of a combined windmill and watermill, one of possibly only 13 in the whole country. The Buckmaster family has owned it or worked it since 1610. It was last worked in about 1925 and was then left as it stood. *"I was born in 1922 and I cant't remember any milling being done at all,"* said the former owner, the late Mr. J Buckmaster Linney. *"The mill is entirely as when they left off,"* he explained. The sails drove two pairs of stones and the waterwheel drove the other two. Although some of the timbers, particularly the floors, are decayed, the mill as a whole is in a remarkably complete state. It was built between 1815 and 1825, probably using machinery from and earlier mill. The drive in the windmill is very old indeed. *"The sails blew off in 1868,"* said Mr. Linney. *"There is an almighty wind from the west."* The loss of the windmill was a major disaster for in later years water power on its own was to prove inadequate. *"They had a portable engine for the last four or five years,"* said Mr.Linney, and there is an external wheel to take the belt drive from an engine to prove it. Mr. Linney hoped that it might be possible to restore this mill to working order but the raising of adequate funds has presented a major hurdle. Nonetheless, of all the derelict mills in the county, this offers the greatest prospect for preserving original machinery *in situ.*

A short distance downstream from Doolittle Mill lies **Bellowes Mill** (NGR SP 981196), a 17th-century three-storey brick building converted to a house many years ago. Happily the millwork, by the famous Suffolk firm of millwrights Whitmore & Binyon, has been largely preserved. The original overshot wheel was replaced in 1901 by a turbine.

At **Leighton Buzzard** (NGR SP 918256) was a large four storey mill. Here the family firm of William Simmons Ltd. ran a milling business until taken over by Rank Hovis McDougall Ltd. in 1964. A modern mill was built on the site using roller mills and a turbine. It was rebuilt in 1912 after a fire the previous year and demolished in 1972/73.

Grange Mill, Heath and Reach (NGR SP 910273) is a small mill built of brick with a weatherboarded section over the space containing the two wheels. The roof is tiled. The mill could date from the 17th century, and the present miller's house, standing to the south-east of the mill, is a good brick house of the late 18th century.

The mill ceased working in 1966. Mr. Robert Bates had worked it for grinding animal feed since 1930. In 1979 he wished to sell it and submitted a planning application for its conversion to a dwelling. This was sympathetic proposal involving few external changes and allowed the retention of the machinery. The mill was *listed* as being of architectural and historic interest. The machinery was not however included in the listing. In June 1979 the mill was sold by auction

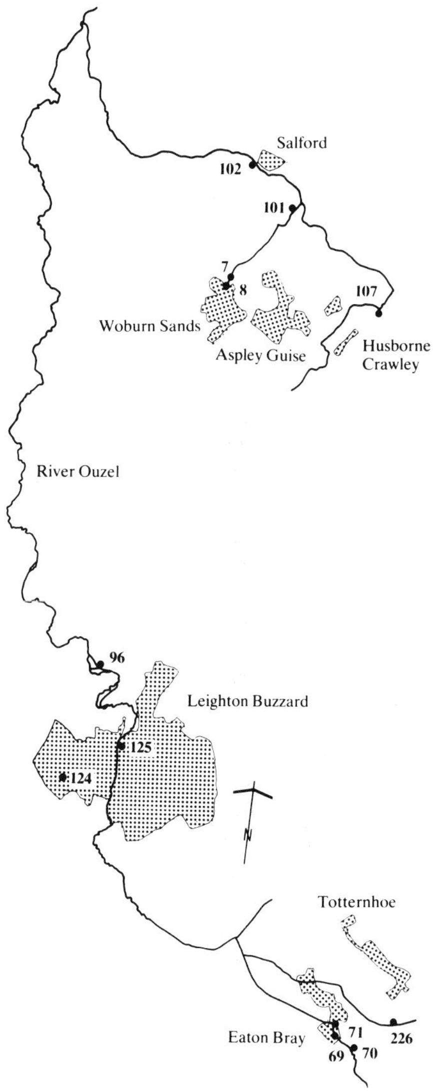

Grange Mill, Heath and Reach, before conversion.

Photo J. K. Major

and a perhaps less sympathetic conversion has been carried out involving additional windows and removal of some of the machinery.

The two wheels were made with sheet iron buckets mounted on timber framing. The shafts in both wheels have been replaced by iron. The west mill had an arrangement whereby the water wheel could be disengaged so that the mill could be turned by a portable steam engine standing outside the mill. Both halves contained two pairs of stones and the associated milling equipment.

Interior of **Grange Mill, Heath and Reach,** before conversion showing vat, horse, hopper, shoe and crook string.

Photo J. K. Major

The Ouzel has a small tributary which rises near Woburn. It flows through the parish of Hulcote and Salford, where there are two watermills, before flowing into the Ouzel near Newport Pagnell. The watermill at **Hulcote** (NGR SP 940381) stands close to the remains of a windmill. It ceased working in 1910, all machinery was removed and the shell converted to a house.

Salford Mill (NGR SP 931393) was kept at work until river works required the removal of the wheel during the Second World War[1] The present structure was built in 1911. It is a small mill with only one set of stones. No doubt the combination of a 16 ft. wheel and an 8 ft. fall gave it more than adequate power. Outside there is a metal bound millstone inscribed 'Hughes & Sons, London'.

[1] **A Thousand Years of Milling**, *Bedfordshire Times*, January 1947

The Flit Mills

The river Flit[2] rises in the grounds of Woburn Abbey. It flows in a generally easterly direction, dropping some 250 ft. in 16 miles and joining the River Ivel at Langford, thus although it is but a small stream its gradient can be harnessed to good purpose. There were six mills on its length, of which Clophill and Maulden had — and Flitwick Mill still has — the more efficient overshot wheels, thus maximising the power from a limited volume of water. All the mills are small and are built in the vernacular tradition. They had to be sited on artificial channels which were made wider and deeper than the main course of the river in order to augment its speed and power.

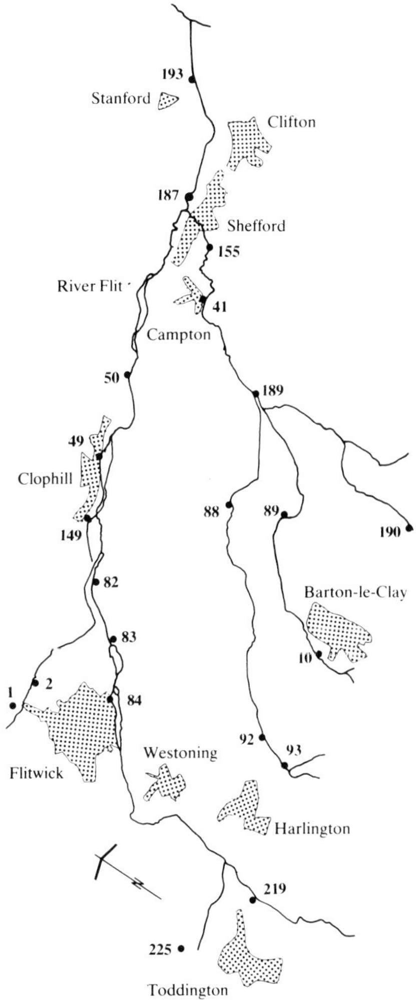

The Park Farm and the estate offices of the Woburn Estates are a fine group of buildings forming two sides of a square. The granary is part of the east range of this and the mill (NGR SP 960332) is part of the granary. The water for the mill is stored in the Leg of Mutton Pond[3]. The leat then runs underground until it discharges on to the wheel by means of an inverted siphon. The overshot wheel is made of iron, 4 ft. wide by 20 ft. diameter. A toothed ring has been fitted to the rim of the wheel and this engages with a spur gear on a lay shaft, which crosses the wheel space from side to side. On the further end of this lay shaft there is a gear drive to a water pump. The other end of the lay shaft carries a gear wheel which engages with a gear wheel on a second lay shaft. This second lay shaft continues until it reaches the hurst frame carrying the millstones. At this point an arrangement of equal bevel gears turns the upright shaft. The great spur wheel on the upright shaft drives two pairs of stones and the upright shaft continues to the crown wheel and clutch for the sack hoist. Auxiliary power was provided by means of a 25 hp Crossley gas engine of 1895.

The surviving machinery, largely of iron, appears to date from the second half of the 19th century. The framing of the first floor suggests that the layout was originally different, with the main shaft on the central axis of the building.

The machinery is designed to drive two pairs of millstones and to operate pumps. The function of the pumps may have been to return water to the mill pond after use. It appears more likely however that the original purpose was to pump the estate water supply, possibly from a deeper well of purer water. For this purpose the water wheel may well have been the principal power source, the gas engine being provided to maintain supplies (and to keep the mill at work) during periods of water shortage. In its later years, when a mains water supply became available, it may have been used to top up the pond above. It seems likely that the present gas engine might have been preceded by a steam engine.

At **Flitwick Mill** (NGR TL 042347) one of the few watermills still at work in the county, the mill pond is embanked some 12 ft. above the

[2] **The River Flit,**, *Lock Gate,* Vol 2 (Local Study by David Falke 1965)

[3] Notes by Students at the Architectural Association; Sites and Monuments Record No. 1689

surrounding fields. Water supply was often a problem particularly in the summer at harvest time when it was most needed. John Goodman, the eighth generation of the Goodman family to run the mill said,

> "If you had a very dry summer you could only be grinding for an hour in the morning and an hour in the afternoon because of the low supply of water,"

The first storey is brick and the upper storey has ochre-painted weatherboarding. There is a luccam (an overhanging structure housing a sack hoist) over the mill door and the roof is slated. The miller's house and stables make an attractive group of buildings.
The mill-work[4] is important in that it is typical of the early 19th century, though it has been repaired at various times since its erection. The iron overshot wheel is 12 ft wide and 10 ft diameter. The main shaft carries a cast iron pit wheel inside the main wall. This engages with a cast iron wallower on the upright shaft. Two pairs of stones are driven by the great spur wheel. A third pair of stones was driven by a steam engine but this has now been removed and is replaced by an electric motor.The wooden upright shaft drives a sack hoist, dressing machines, and a roller crusher from the crown wheel on the first floor.

There are no visible remains of **Greenfield Mill** (NGR TL 054534) but at **Maulden** (NGR TL 073373) there is a fine example of a mill and miller's house under one roof though the mill itself has been gutted. It formerly had an internal overshot wheel of ten foot diameter driving two pairs of stones. The next mill, at **Clophill,** (NGR TL 086378) is an urbane Georgian building in chequered brickwork. It has not been used sincc before the last war; the wheel and most of the machinery is missing. **Shefford Mill** (NGR TL 150395) which augmented the ineffective windmill (referred to in Chapter 5) was destroyed by a flood in 1959. The final mill on the Flit was at **Stanford** (NGR TL 171409) but is believed to have been demolished in about 1900.

A brief mention must be made of a tributary of the river Flit. Known as *the Running Waters* it is *"a very important little stream"*, according to Mr Charles Scott. There were two mills on it. One was **Doolittle Mill**, Ampthill, where Charles Scott's brother was the miller for some twenty-three years. It is believed to have fallen out of use at about the time of the First World War. It still stands to the west of the spot, remembered by Charles Scott, where the stream crossed the Ampthill-Flitwick road. *"And they used to come down there forty or fifty at a time and they would be paddling in the water as it came across the road and they would have a grand time."* One wonders if the children of Redborne School know what they are missing!

"Down a little further there was another water mill." This was the one to which Charles Scott as a child accompanied the last load of corn in about 1886. Its remains were discovered in 1982 by Ampthill and District Archaelogical and Historical Society. These suggested

[4] **The Windmills and Watermills of Bedfordshire** — *A Survey* 1967, J Kenneth Major

????

Volary Account

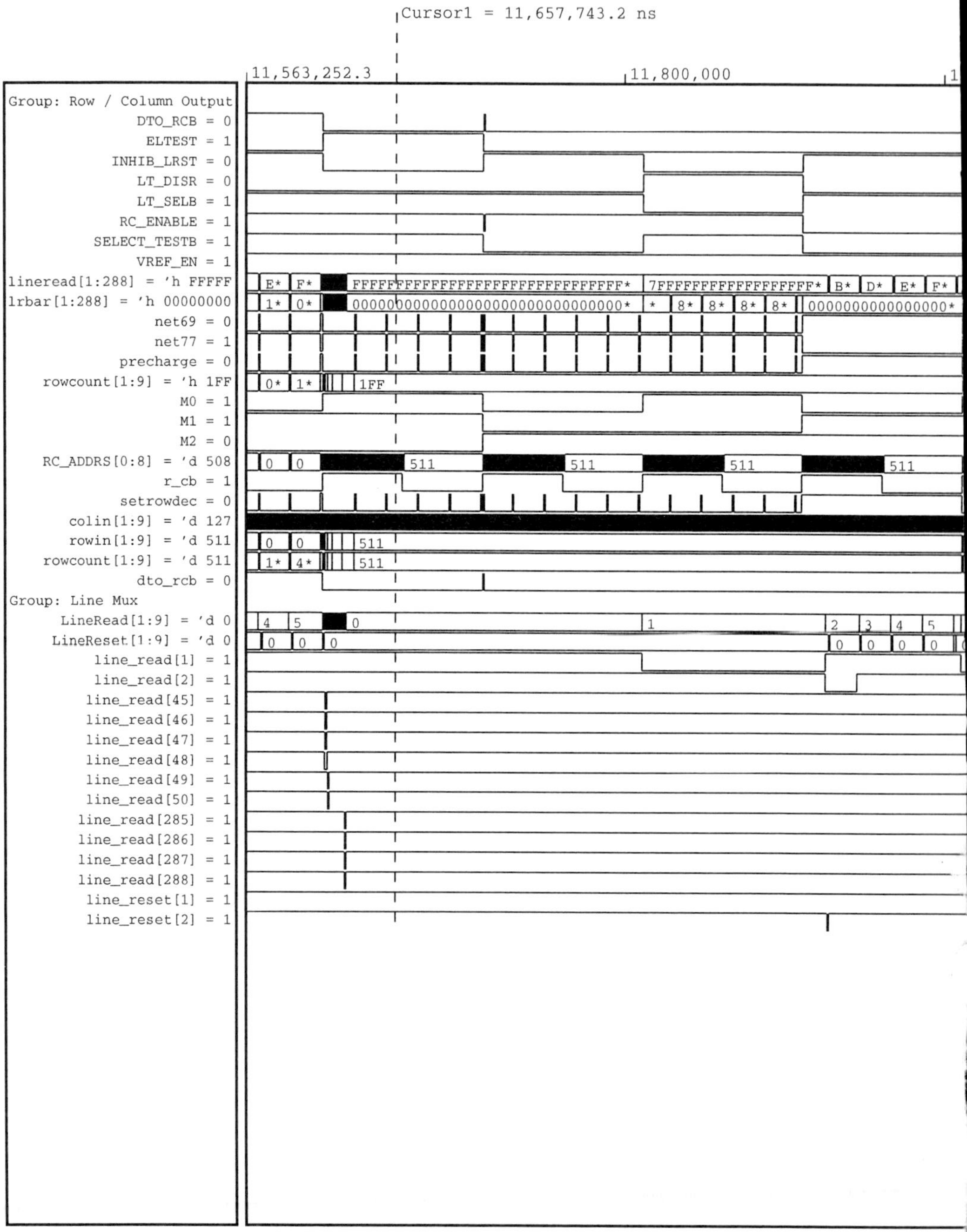

that it probably had a low breast-shot wheel of about 15 ft diameter and about 6 ft width.

The Campton Brook and Barton Brook, neither of which is more than three miles long, flow into the River Flit at Shefford. Three mills still stand on them. **Shillington Old Mill** (NGR TL 119356) is a two storey yellow brick building with a luccam. It has not been used for milling for many years. **Campton Mill** (NGR TL 131378) is a weatherboarded timber structure on a yellow brick ground floor. The millwork is thought to be by Course & Son but the structure is rapidly deteriorating and is in a dangerous condition. However **Harlington Grange Mill** (NGR TL 055312), a 17th century half timbered building, is in sound condition though devoid of machinery.

Barton Mill was probably built in 1852, the date cast on the surviving external waterwheel. In 1974 Mr. L. W. Vass restored it,

> "If we hadn't done something, it would have collapsed. We replaced any of the damaged framework, reclad it entirely and also did some repairs to the roof. Most of the machinery is there but it is not absolutely complete. The gear for driving it is complete. There is one pair of millstones which is complete and one that has only got the bedstone. One set was removed and sold (presumably the more valuable French burrs). It can't be put back into working order: the main reason being the stream which supplies it passed under the Barton-Sharpenhoe wall. There used to be a hump bridge. The County Council wanted to get rid of that arch. And so, as the mill was not in use, they came to some arrangement to deepen the bed of the stream so that it took away the hump-backed bridge. That meant that there was no longer access to the water coming to the mill head."

Barton Mill being restored in 1974.
Beds. C.C. Photographic Unit

The Ivel Mills

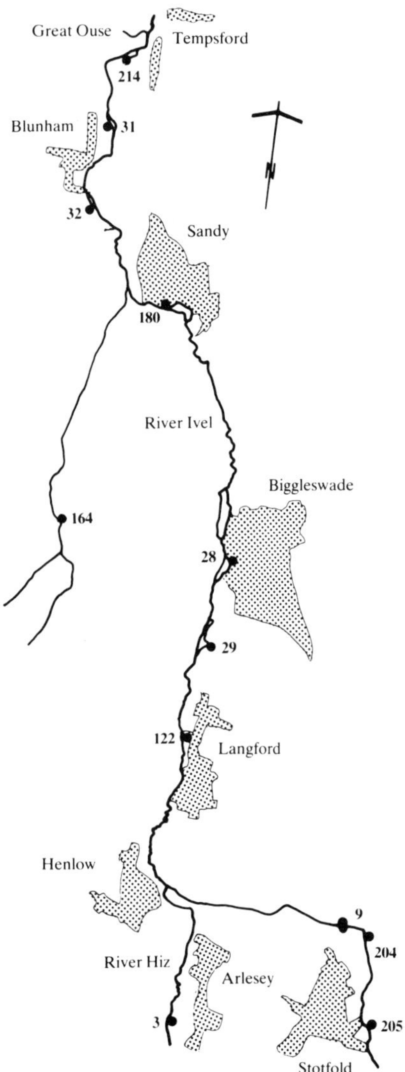

The River Ivel is 20 miles long from its source near Baldock to its confluence with the Great Ouse at Tempsford. It carries a substantial volume of water particularly below its junction with the Flit. *"It is a very good river,"* said John Jordan. *"The water flow is always good, even in summer."* Nevertheless most millers found it necessary to use auxiliary power to make the mills work more continuously. Its development as a short-lived navigation encouraged the construction in the middle of the 19th century of several large *industrial* mills where there were weirs or locks.

The access to other parts of the country by water or rail made these mills very successful and several were modernised, turbines taking place of wheels and roller crushers substituting for stones. In this way they were able to compete for many years with more modern mills.

Sandy Mill and **South Mills**, Blunham, have both been demolished in recent years but otherwise the Ivel mills have survived. The river enters the county just south of **Stotfold** where the **Old Mill** (NGR TL 224367) forms an impressive complex with its tall chimney and 18th century mill house. The ground floor of the mill itself is built of white gault bricks with two further storeys clad in weatherboarding under a slated roof. Although the steam engine is now missing the boiler house is still standing. Also known as *Randall's Mill,* it only ceased working 15 years ago and the millwork is believed to be still *in situ.* It was made by the famous Suffolk millwrights Whitmore & Binyon. The large overshot waterwheel was 10 ft. in diameter by 16 ft. wide and made entirely of iron. It was removed for salvage during the Second World War and replaced by a turbine. There is an iron hurst frame carrying two pairs of peakstones and two pairs of French stones.

A mile further down the river is **Taylor's Mill** (NGR TL 221383) which is a three-storey vernacular building dating from the mid-18th century, incorporating two cottages under its long roof. It is something of an anachronism amongst the more modern industrial mills of the Ivel. Unfortunately all the machinery has gone and replacement of some timbers and flooring makes it difficult to estimate even how many pairs of stones there were. It seems that the mill became disused during the early years of this century. There are marks on the brickwork from the circumferences of the waterwheel and pitwheel. A turbine has been installed in the wheel chamber, taking advantage of the 5 ft. fall to generate electricity. Supplied by Armfield Hydraulic Engineering, it was originally installed at Radwell Mill (just over the Herts border) in 1929 and moved to Taylor's Mill in the early 1950s. The last owner provided some additional support for the decaying timber and re-roofed the building in the 'fifties. The luccam is missing and a bay over the headrace is believed to have been demolished.

Only a few yards further down the river is **Astwick Mill** (NGR TL 117285) a fine structure built in 1847 as a combined water and steam-

Sandy Mill, demolished in 1977.
Photo J. K. Major

driven mill. Kenneth Major[4] described it in the following terms,

> "The body of the mill is three-storeyed on the upstream side and four-storeyed on the downstream side. The walls are built of white gault bricks and have large circular-headed panels forming a decorative treatment on each face. The boilerhouse with its decorative chimney stands on the other side. The roof is slated. On the north side of the mill there is an earlier building which may have been part of a previous mill.
>
> The millwork of Astwick Mill is important, for it is a fairly complete example of the millwrighting of the mid-nineteenth century. The overshot water wheel is made of iron and is 16 ft. wide by 16 ft. in diameter. Inside the wall of the mill on the main shaft, there is a 9 ft. diameter spur gear which engages with a 5 ft. diameter secondary gear on a lay shaft. This lay shaft drove the pit wheel which stands inside the octagonal hurst frame. The wallower engages with this pit wheel and turns the upright shaft of the mill. The great spur wheel on the upright shaft drives the 8 pairs of stones. The hurst frame is entirely built of iron with classical detailing on the columns. A clutch exists which engaged the steam engine but all the steam plant is now missing. There is a crown wheel for the other machinery which has now gone from the upper floors."

Langford Mill (NGR TL 183413) is similarly a large *industrial* building but stripped of its machinery and now given over to the production of seed grain for Dalgety Seeds.

A mile further down the Ivel brings us to Bedfordshire's sole surviving corn mill — **Holme Mills,** Biggleswade (NGR TL 186430). Here Jordan's wholemeal flour is still ground on rollers powered by the 1890 Gilbert & Gilkes turbine. The appearance of the mill building has been much changed by the addition of a further floor. To appreciate what it used to look like you have to look at the

[4] **The Windmills and Watermills of Bedfordshire** — A Survey 1967, J Kenneth Major

picture on the flour bags. John Jordan explained the hair-raising experience that had led him to make the change,

> "When I first came here there was an ordinary pitched roof. I used to have to go out of a ventilator and luckily I used to go on a rope and clean the guttering. You get a lot of dust blowing out on the roof and it was slippery. I was right on the end of the rope and I slid off the roof and over the edge of the guttering and I said, *'if ever I get back up there I am going to put a flat roof on here'*. Well, I did get back and I did put a flat roof on!"

The shell of **Biggleswade Mills** (NGR NL 187444) is all that remains of the headquarters from which Henry Franklin conducted his epic struggles with William Jordan, as recounted in Chapter III. The fine three-storey *industrial* building was badly damaged by fire in August 1945. It was then used for storage purposes until sold by Dalgety Agriculture Ltd. in 1979. In 1982 it was converted to six flats. As the chances of the building ever being used for milling again were minimal it seems a most expeditious way of preserving this fine building.

South Mills, Blunham, before demolition in 1966.

Beds. C.C. Photographic Unit

No such happy story can be told of **Sandy Mill** (NGR TL 17488). This was a further fine late-Victorian three-storey Mill building with a two-storey wheel house. It was demolished in 1977.

The next mill, **South Mills,** Blunham (NGR 154504) was also demolished in recent years. This fine building, designed by John Usher the famous Bedford architect, was used for milling fertiliser from bones and slaughterhouse waste. Jim Battison recalled,

> "We used to go down to see them cut the horses up. Flesh was sent up to London. Every part was used."

Kenneth Major[5] described the building in some detail before it was demolished,

> "The mill is built of white gault bricks which have mellowed very attractively to make a fine building in the industrial tradition. The chimney is half buried in the wall of the furnace bay, and its present height is some fifteen feet lower than its original height. The important feature of the chimney is the very fine podium on which it stands. The warehouse to the east of the main building is a well-built structure with panelled brick walls. The roof of this warehouse runs in two bays from east to west. Apart from the excellence of the brick detailing of the buildings, the important feature of the structure is the system of cast-iron columns and floor joists. Another interesting feature of the mill is the spiral staircase. This is the made of solid elm treads and is generated within a square tubular space.
>
> Mechanically, the only items of equipment remaining from the oil cake mill of the first user, are the two wheels and the large edge-runner stones which are now propped against the warehouse wall. Another piece of equipment from this period is the great oven — now a paint store — where the oil seeds were heated to release the oil before they were ground under the edge-runner stones.
>
> The outer of the two wheels did not serve the mill, but is believed to have been connected to a pump which stood still further to the north of the mill. This wheel is the same as the main wheel, but has only two sets of spokes supporting its buckets. The main wheel consists of two bays— that is it had three sets of spokes. The spokes and rims are made of cast iron, with wooden supports to the buckets. The present buckets are of sheet iron though they will have been made of wood when they were first installed. The hubs and axle are cast iron, and there are six spokes mounted in two semicircular castings on the axles. A further interesting feature of these wheels is that the bucket supports are wedged and not bolted into place.
>
> The equipment remaining from the bone-crushing plant consists of the drive system from the main wheel and the line shafting to the various pieces of the machinery. On the ground floor there is a solidly built bone-crushing machine, and at a high level above the floor there is a dessicator for further dressing of the bone flour. On the first floor there is the sack pulley which is controlled by a belt from the shafting below. This belt is actuated by increasing the pressure on the belt by means of a lever and an idling pulley. The sack hoist operated through the sack traps, and also through the pulley on a Heath Robinson crane beam on the top floor. On the fourth floor there is a dressing machine for the particles of bone and the bone flour. This dressing machine was connected to a dust extraction room on the top floor which has lost its mechanical equipment."

5 *Journal of Industrial Archaeology*, Vol III, No. 3, August 1966.

Blunham Mills (NGR TL 158530) is a fine building, spanning a central mill-race where the two waterwheels are housed in the middle of the building. It is a timber-framed *vernacular* building above a brick ground floor. It is clad with weatherboarding under a tiled roof with modern skylights. When Rex Wailes surveyed the building in 1962 he reported that it was *a complete derelict* and likely to collapse. The present owner has given it a new roof using the original tiles. The danger of further collapse has therefore been held in check. There are the remains of two wooden wheels. Each wheel drove a separate set of gearing at opposite ends of the mill. That on the right (facing downstream) was the larger wheel and drove three pairs of stones through a wooden upright shaft; that on the left drove two pairs through an iron shaft. The mill work is clearly of the *pre-industrial* era with no indication of the identity of the original millwrights. At the time of his survey Rex Wailes estimated that the machinery was about 100 years old. Most of the machinery is still *in situ* and not beyond eventual restoration.

The last mill is at **Tempsford** (NGR TL 158530). It was converted to a saw mill after the fire referred to in Chapter IV. In 1824 it had been described as a water corn mill with two water wheels and two pairs of stones. Below the mill the waters of the Ivel mingle with those of the Ouse. We turn next to the mills powered by the *Silent Tide* of that great river.

Blunham Corn Mill in the early years of the century.

Photo supplied by R. Izzard

The Mills on the Great Ouse

The Great Ouse is already a mature river flowing in a well-defined flood plain. The river gradient is flat but the volume of water is large and constant. These conditions made it impossible to use the efficient overshot waterwheels, because of the lack of vertical gradient. In fact, this was unncecessary, because there was always adequate water to work the less efficient undershot and breast-shot wheels over the correspondingly longer periods. The mills display a mixture of the rural vernacular and industrial styles of architecture. There are mills or known mill sites at Turvey, Harrold, Odell, Sharnbrook (3), Milton Ernest, Bromham, Kempston, Bedford (2), Castle Mills, Cardington, Willington and Great Barford. Other possible sites were at Carlton, Felmersham, Bletsoe, Stevington, Oakley, Clapham, Box End and at Ford Mill near Queens Park in Bedford.

Over the years the water authorities have sought to lower the level of the river with a view to converting the water meadows to productive arable land. On many stretches of the river the level is now 18 inches lower than before and as a result it is more difficult to operate water wheels. Furthermore the straightening of the channel has meant the loss of much of the former rural character of the river.

Turvey Mill (NGR SP 938523), the first in the county, is now only used for light industry. Nevertheless the owners have maintained it well and kept its original appearance. The view of it from the 15th-century bridge is charming. The white newly painted weatherboarding of the luccam contrasts with the honey-coloured lime-stone.

The setting of the next mill at **Harrold** (NGR SP 950556) is equally pleasing amidst the willows. The mill itself has been converted to a house which retains little of the original character.

Odell Mill (NGR SP 967577), too, was converted to a house in the early 1950s. Here the waterwheel has been retained but the works to the river have meant that the mill stream is dry and no water now passes through the overflow sluice, the work of Thomas Course & Son of Bedford.

Nothing remains of **Felmersham Mill** or of **Pinch Mill** at Sharnbrook. Fortunately the late Frank Hipwell[6] documented much of the history of the latter,

> "Well away from the beaten track, Pinch Mills, near the border dividing Sharnbrook from Felmersham Parish, only exists now as an entangled site. Having been unoccupied and at the mercy of the elements for many years, demolition of the ancient buildings was determined on in 1880 and a thorough job done, not one stone being left upon another above ground, though a glimpse of foundations can be seen when the river is low, and a quantity of rubble still lies in the stream itself.
>
> The original layout of the ancient mill with its house and premises is not easily traced, but the main structure spanned the river and its wheel

6 **The Mills of Sharnbrook**, Frank Hipwell, *The Sharnbrook Review* 1949

The massive cast iron pit-wheel at **South Mills Blunham.** On the right is the sluice gear for controlling the flow of water to the waterwheels which were located the other side of the wall.

Beds C. C. Photographic Unit

The Great Ouse Mills

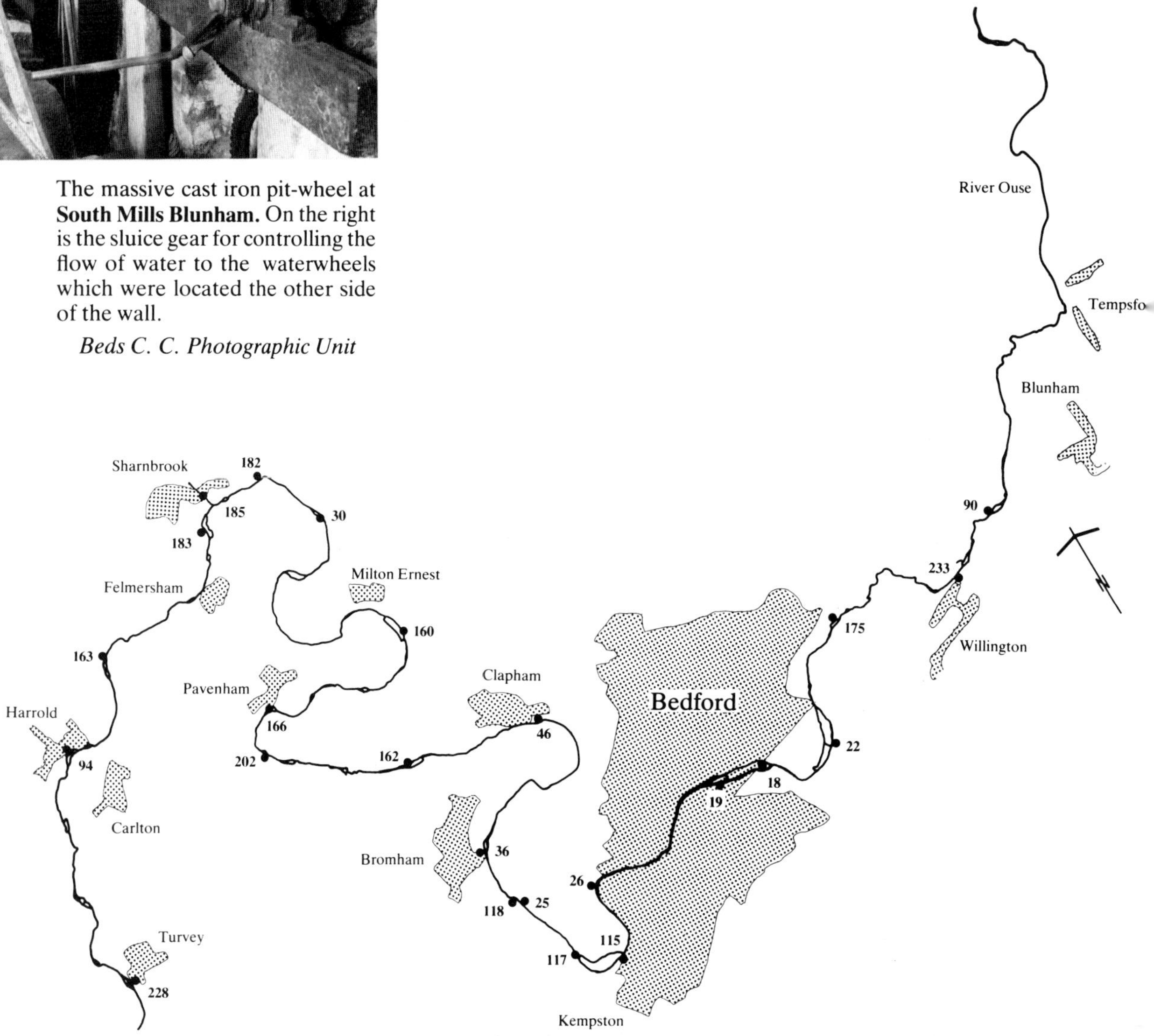

house was furnished with three gates. These discharged into the now silted-up and rusty millpit, one delivering direct to an undershot wheel, while others, less frequently raised, were there to give passage to the rush of water in time of flood, which also was provided against by slip-ways still in evidence. A *cut* to allow pleasure boats to pass from Sharnbrook water to the Felmersham reaches was made shortly after the demolition, for the main waterway was left too blocked to admit of such use, and several field dykes flow into it. The effect of these numerous channels is to create a miniature archipelago, its islands smothered in a dense jungle of undergrowth among thriving trees and old fallen trunks.

It would seem that the Mill could only have been a small one, containing two or at most three pairs of stones, since the *fall* and consequently the power obtainable from the river at this point, is very limited. A ford is said to have existed near-to at one time, and of course when the mill was in being its bridge made a short-cut between the two hamlets."

In contrast to the limited fall at Pinch Mill, that at **Stoke Mills** (NGR TL 010591) 8 ft., is the biggest on the river. This was one of the biggest and most successful of the corn mills in the county. It milled flour until 29th June 1969 when it was *silenced*. It is a fine tall brick building, under a slate roof with deep eaves adorned by drop finials. The present building dates from about 1892. The design was by Usher & Anthony, architects and surveyors of Bedford. The turbine is still in place but all the machinery including the banks of roller mills, dressers and purifiers has, as recounted in Chapter III, been removed.

An early photo of **Bromham Mill.**
Photo Bedford Museum

Nothing remains of **Radwell Mill** which probably stood near where the railway bridge is today. Still less is known of a possible fulling mill at **Stevington**. This is a pity as no other such mill is known to have existed in the county.

The shell of **Milton Ernest Mill** (NGR TL 019550), once part of the Milton Ernest Hall Estate, is clearly visible from the A6. Like the hall, it was designed by William Butterfield whose most famous building was probably Keble College, Oxford. The utilitarian mill was treated with much less flamboyance.

The site of **Clapham Mill** gives rise to a mystery. At Domesday it commanded a rental of no less than 40 shillings, one of the highest on the Ouse. Only the rental of Bromham exceeded this figure, *by 100 eels*. During 1981 the dredging of the Ouse brought to light some masonry which is possibly the remains of the mill. But there are no other remains nor indeed any records of the mill. Nor is there any indication of when it went out of use. The mystery of why such an apparently prosperous mill has disappeared without trace therefore remains.

The payment of rent in eels is an interesting sidelight. The twenty-five mills which paid rent partly in cash and partly in eels were all situated on the Ouse. Eels were clearly a profitable sideline. They were sold in *sticks* of twenty-five and in the time of Edward I (1272) the price of eels was fixed at two pence per stick. They were clearly a valuable commodity until recent times. An eel trap is still to be seen in the head race of Bromham Mill. Jim Battison remembered a similar trap at Blunham in the early years of the century. He said,

> "Eels used to be an industry, they used to send the eels up to London. I remember an old chap was had up for poaching eels."

Bromham Mill (NGR TL 011580) is a two-storey red brick building in rural vernacular with some style half-timbering. The mill has been renovated by the Bedfordshire County Council. The building has stones dated 1695 and 1722 but the present structure probably dates from the late 18th or early 19th centuries. It was still working, by electricity, when the County Council bought it in 1974. One wheel, manufactured in 1908 by Anstee of Kempston, is in working order and drives two pairs of stones. The other wheel which fell to pieces in 1939 used to drive three pairs of stones.

Kempston Mill (NGR TL 022477) was a particularly imposing five-story building which was destroyed by fire in 1969. Little remains apart from some of the ground floor stonework and the waterwheel pit complete with sluice gear. This is now dry following the major changes to the course of the river in recent years.

The mill was little changed for hundreds of years until the turn of the century, when a roller plant was installed in a large five-storey timber building joined on to the old mill. At the same time an auxiliary steam engine was added with a tall brick chimney. A Swiss turbine was installed in about 1920 in place of one of the wheels. John Clover recalled, "*The mill ran continuously through two World Wars, particularly during the latter.*" In 1939 most of the milling capacity of the country was centred at the main ports and much was destroyed in the blitz. The importance of the country mill was vital, and Kempston Mill ran without respite. In 1952 the mill gave up flour production and turned to animal feed production. By this time, water played a minor role in power requirement, and the main source was electricity.

A considerable business was built up, resulting in a complex of buildings for the different products, and equipment was installed to carry out all the millwrighting and engineering involved. In 1968 it was decided to give up milling and to concentrate on the engineering side; a business was built up in reconditioned milling machinery, for which the present set-up was ideal.

There is little evidence left of any further mills on the Ouse in Bedfordshire. **Duck Mill** in Bedford was an 18th-century watermill which was demolished in 1891 to make way for the weir. This was part of the Town Council's "programme for improving the amenities in Bedford and for preventing the flooding which was in former years such a regular and unwelcome visitant in the adjacent meadows and low-lying part of the town". No technical details are known but the site is familiar to most Bedfordians as **Duck Mill Bridge**. (NGR TL 055494).

Kempston Mill burnt down in 1969.

Photo supplied by J. Clover

Newnham Mill, Bedford.

A watercolour by Bradford Rudge, 1850, from "Old Bedford" by C. F. Farrar, published 1926

Duck Mill, Bedford, from the east.
From "Old Bedford"

The site at **Cardington Mill** (NGR TL 079489) is also familiar to many, even though the building was demolished in 1936. This mill was built to the designs of the great engineer John Smeaton. The chequered history of the mill is recorded on a stone on the site—

> THIS STONE MARKS THE SITE OF
> CARDINGTON MILL
> REBUILT BY S WHITBREAD ESQ. 1786
> J SMEATON ENGINEER
> WAS DESTROYED BY FIRE DEC 1823
> REBUILT BY W H WHITBREAD ESQ. 1824
> AGAIN DESTROYED BY FIRE AUGUST 1840
> AND REBUILT THE SAME YEAR BY
> W H WHITBREAD ESQ. AND DEMOLISHED
> BY THE BEDFORD CORPORATION NOV 1936

The shell of **Castle Mills** (NGR TL 092509) still stands despite being burnt down in 1935. It had a large iron and steel waterwheel 16 ft. in diameter and 10 ft. wide.

There was a large corn mill at **Willington** (NGR TL 111503) boasting two waterwheels and three pairs of stones. The mill itself was demolished in 1850 but the site could be easily identified by sluices, weir and lock until it was completely obliterated in recent years by the works for restoring the navigation.

These same works did, however, reveal a previously unknown site at **Great Barford**. When the navigation was being restored in 1974 the base of a mill was found with wheel pit and a possible axle bearing for a waterwheel. The base appeared to be made out of brick with wooden beams and floor boards. There is no documentary evidence of the mill and it must presumably have been demolished before the lock was built, as the mill would have blocked the navigation.

Cardington Mill. Demolished in 1936.

Photo Bedford Museum

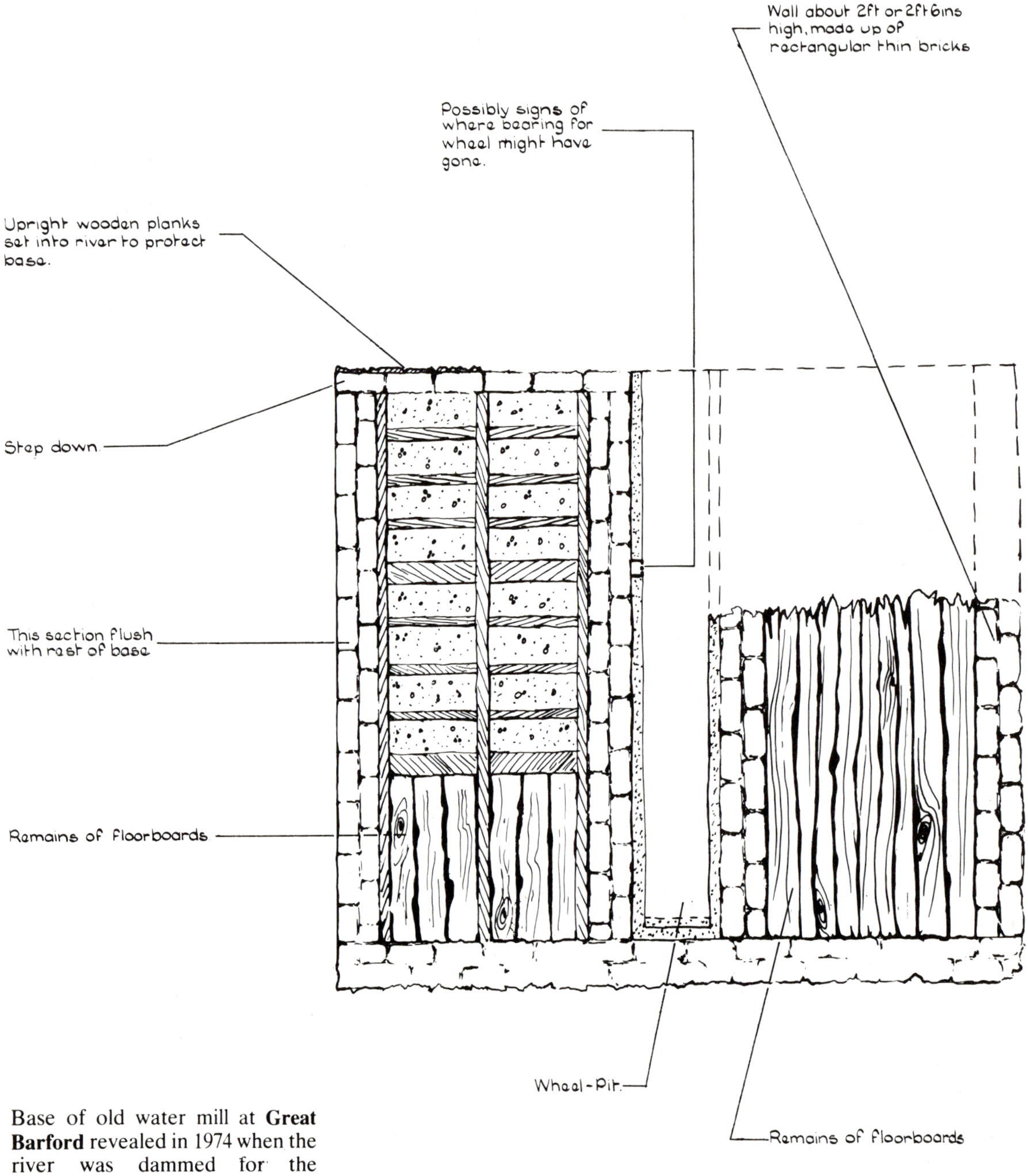

Base of old water mill at **Great Barford** revealed in 1974 when the river was dammed for the restoration of navigation of the River Great Ouse.

It measured approximately 30 ft in each direction.

The Lea Mills

The remaining river in the county which supported watermills is related to London and the Thames. The Lea is a shallow and sluggish stream rising in Leagrave Marsh. Nowadays the first mill is at Hyde but in Saxon times there were no less than six between it and the source. William Austin, the Edwardian historian, identified these as Leagrave Mill, Limbury Mill, the North Mill, Luton, the Church Mill, The Brache Mill and Stapleford Mill. The actual sites could not always be precisely located and there are no physical remains of any of them.

The watermill at Hyde is a brick building mostly dating from 1835 and is still occasionally worked. The county boundary cuts through the middle of this interesting complex of buildings which are built of local brick with slate roofs. The mill itself is in Hertfordshire with the abutting millhouse lying in Bedfordshire. It is a fine example of a mid-19th-century millhouse reflecting the miller's prosperity.

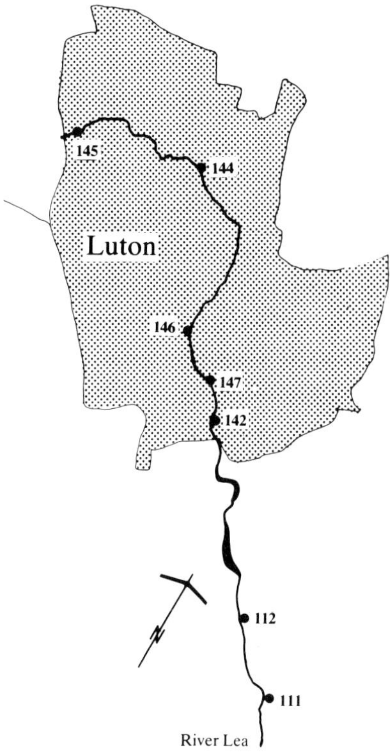

Hyde End Mill. *J K Major*

Above

The scene at **Shefford** in about 1920 showing the ineffective but still surviving mill tower on the right and the watermill which was destroyed by floods on the left.

Campton Parish Council

Facing page-top

The mill-head at **Flitwick water mill** in about 1910 showing a long forgotten boiler chimney.

County Record Office Andrew Underwood Collection

Facing page-bottom

An early photo of **Maulden water mill** — a good sample of a mill and mill house under one roof.

County Record Office Andrew Underwood Collection

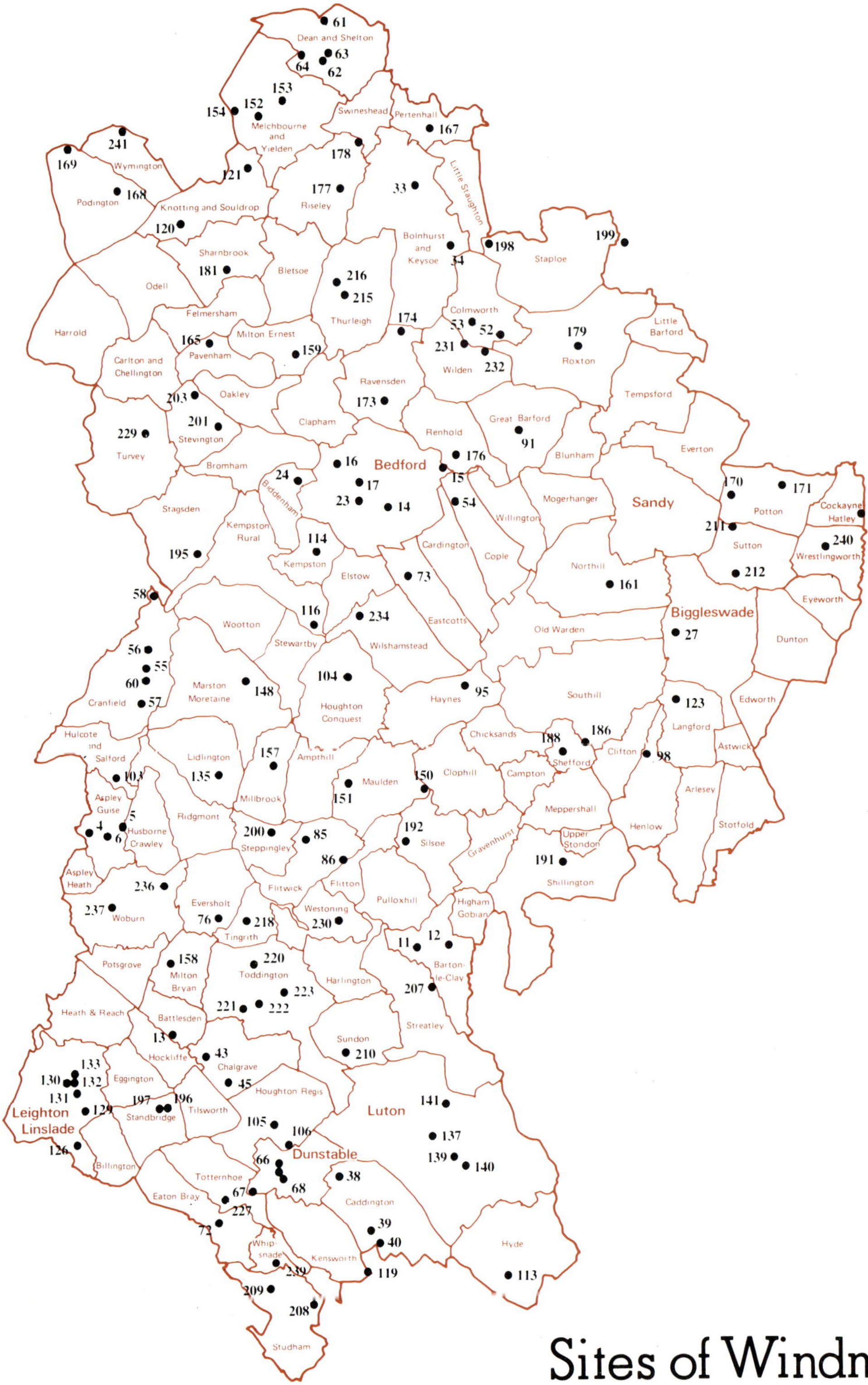

Sites of Windmills

Gazetteer of Mill Sites

	Parish & NGR	SMR	Notes
Ampthill			
1	TL 027 365	1005	Doolittle Mill (watermill building)
2	TL 036 364	2618	watermill site
Arlesey			
3	TL 189 363	5201	watermill site
Aspley Guise			
4	SP 930 364	2487	windmill site
5	SP 949 361	8697	windmill site
6	SP 940 359	10110	possible windmill site
7	SP 930 364	2636	watermill site
8	SP 929 363	10114	watermill site
Astwick			
9	TL 217 384	1010	existing watermill
Barton-in-the-Clay			
10	TL 077 312	1015	existing watermill
11	TL 075 311	3198	windmill site
12	TL 087 312	2619	watermill site
Battlesden			
13	SP 966 276	3196	windmill site
Bedford			
14	TL 065 494	0291	windmill site
15	TL 089 512	3178	windmill site
16	TL 044 512	3193	windmill site
17	TL 054 504	3194	windmill site
18	TL 065 493	6601	site of Newnham Mills (water)
19	TL 055 495	2620	site of Duck Mill (water)
20	TL 061 505	3195	watermill site
21	TL 052 489	4534	existing former steam mills
22	TL 079 489	2623	site of Cardington Mill (water)
23	TL 053 498	12138	wind and steam mill site
Biddenham			
24	TL 025 506	3192	windmill site
25	TL 012 496	11513	watermill site
26	TL 026 490	11514	watermill site
Biggleswade			
27	TL 189 440	0928	site of Franklin's tower mill
28	TL 188 444	1690	converted watermill
29	TL 185 430	2499	Holme Mills (water)
Bletsoe			
30	TL 017 577	12132	watermill site
Blunham			
31	TL 157 518	1025	existing watermill
32	TL 154 504	2486	South Mills (water and steam) rebuilt
Bolnhurst and Keysoe			
33	TL 077 627	0930	site of Keysoe windmill
34	TL 094 602	2508	windmill site
35	—	7800	possible watermill site
Bromham			
36	TL 011 507	1029	existing watermill
37	TL 002 510	2621	watermill site
Caddington			
38	TL 039 214	2882	windmill site
39	TL 052 192	3190	windmill site
40	TL 056 186	3191	windmill site
Campton			
41	TL 130 378	5629	existing watermill
Carlton			
42	—	12134	watermill site
Chalgrave			
43	SP 983 268	3176	windmill site
44	—	11806	windmill site
45	SP 992 257	11807	windmill site
Clapham			
46	TL 032 523	1199	watermill site
47	—	6602	windmill site
Clifton			
48	—	12136	watermill site
Clophill			
49	TL 086 379	2610	existing watermill
50	TL 106 383	9130	site of Beadlow Mill (water)
Cockayne Hatley			
51	TL 270 490	1616	windmill site
Colmworth			
52	TL 113 574	2894	windmill site
53	TL 102 570	4501	possible windmill site
Cople			
54	TL 094 496	3188	windmill site
Cranfield			
55	SP 960 429	0933	site of tower mill
56	SP 960 437	2530	windmill site
57	SP 946 414	3186	windmill site
58	SP 964 460	3187	windmill site
59	—	10650	site of horse-driven malt mill
60	SP 959 422	10651	windmill site
Dean and Shelton			
61	TL 041 698	0926	remains of post mill
62	TL 040 681	0934	ruined tower mill
63	TL 043 083	8137	windmill site
64	TL 068 692	8207	windmill site
65	TL 030 683	8210	possible windmill site
Dunstable			
66	TL 013 221	0129	windmill site
67	TL 001 208	3185	windmill site
68	TL 015 220	2609	existing converted tower mill

Parish & NGR	SMR	Notes
Eaton Bray		
69 SP 970 198	1179	Two Counties Mill (water)
70 SP 981 196	1687	Bellowes Mill (water)
71 SP 976 200	2622	watermill site
72 SP 986 196	2827	windmill site
Eastcotts		
73 TL 073 467	3189	windmill site, Harrowden
Elstow		
74 —	6603	windmill site
Eversholt		
75 SP 988 322	2829	site of horse-driven corn mill
76 SP 988 325	3182	windmill site
77 SP 991 334	9432	possible watermill site
78 SP 982 325	9608	steam mill building
Eyeworth		
79 —	3181	possible windmill site
Felmersham		
80 —	12133	watermill site—Felmersham
81 —	12135	watermill site—Radwell
Flitton		
82 TL 063 365	1850	watermill site
83 TL 054 354	2624	site of Greenfield watermill
Flitwick		
84 TL 042 348	1061	existing watermill
85 TL 027 356	3179	site of smock mill
86 TL 043 348	—	possible windmill site
Gravenhurst		
87 —	6610	possible windmill site
88 TL 095 348	7788	site of Whitehall watermill
89 TL 100 336	9783	site of Sangerel watermill
Great Barford		
90 TL 128 510	2608	watermill site
91 TL 122 526	3200	windmill site
Harlington		
92 TL 054 312	2607	Grange Mill (water)
93 TL 051 303	9708	possible watermill site
Harrold		
94 SP 951 566	1067	converted watermill
Haynes		
95 TL 098 420	3177	windmill site
Heath and Reach		
96 SP 910 273	1002	Converted watermill, Grange Mill
97 —	3137	windmill site
Henlow		
98 TL 175 390	0932	base of smock mill
99 TL 175 389	7324	existing former steam mill
Higham Gobion		
100 —	6605	windmill site
Hulcote and Salford		
101 SP 943 382	0935	converted watermill
102 SP 931 393	2613	existing watermill
103 SP 948 382	2835	base of tower mill
Houghton Conquest		
104 TL 047 423	0936	ruined tower mill
Houghton Regis		
105 TL 012 238	0937	base of tower mill
106 TL 019 229	3175	windmill site
Husborne Crawley		
107 SP 965 361	2626	existing watermill buildings and site of clay pug mill
108 —	8698	site of Horsepool Mill (water)
109 —	8699	site of Ypewelle Mill (water)
110 —	8704	site of Nichols Mill (water)
Hyde		
111 TL 133 170	1691	existing watermill
112 TL 122 181	2627	watermill site
113 TL 112 171	3169	windmill site
Kempston		
114 TL 034 477	3174	windmill site, Up End
115 TL 023 476	1081	remains of Kempston watermill
116 TL 034 447	3571	former windmill site, Kempston Hardwick
117 TL 016 482	8117	possible watermill site
118 TL 011 496	8131	watermill site, Box End
Kensworth		
119 TL 053 174	3173	windmill site
Knotting and Souldrop		
120 SP 978 612	1332	windmill site
121 TL 008 635	3172	windmill site
Langford		
122 TL 184 414	2628	Langford watermill
123 TL 187 414	3171	windmill site
Leighton Linslade		
124 SP 905 246	2616	Southcott watermill
125 SP 918 256	2629	site of Leighton mills
126 SP 923 230	3289	windmill site
127 SP 916 257	11010	steam mill
128 —	11020	windmill site
129 SP 928 243	11051	windmill site
130 SP 920 256	11052	windmil site
131 SP 925 253	11053	windmill site
132 SP 924 255	11054	windmill site
133 SP 924 259	11055	windmill site
134 SP 922 258	11093	steam mill
Lidlington		
135 SP 989 385	3170	windmill site
Little Barford		
136 —	12137	watermill site
Luton		
137 TL 080 231	0179	site of Biscot smock mill
138 TL 091 215	2988	site of corn and saw mills
139 TL 090 222	3166	windmill site
140 TL 094 219	3167	windmill site
141 TL 086 245	3168	
142 TL 100 209	8242	site of Brache Mill (water)
143 TL 023 214	10528	site of Gt. Northern steam mill
144 TL 075 244	10814	site of Limbury Mill (water)
145 TL 052 238	10817	site of Leagrave Mill (water)
146 TL 089 218	10818	site of North Mill (water)
147 TL 095 214	10819	site of Abbey Mill (water)
Marston Moreteyne		
148 TL 004 423	3165	windmill site
Maulden		
149 TL 073 373	2615	existing watermill
150 TL 081 378	2963	windmill site
151 TL 045 381	6626	windmill site

Parish & NGR	SMR	Notes
Melchbourne and Yielden		
152 TL 011 657	3134	windmill site
153 TL 021 663	3164	windmill site
154 SP 000 660	8304	windmill site
Meppershall		
155 TL 141 383	5448	possible site of Polehanger Mill (water)
Milbrook		
156 TL 012 386	2630	watermill site
157 TL 013 387	3163	windmill site
Milton Bryan		
158 SP 966 306	3162	windmill site
Milton Ernest		
159 TL 027 555	0908	windmill site
160 TL 019 549	2614	existing watermill building
Northill		
161 TL 160 462	3161	windmill site
Oakley		
162 TL 008 530	1320	watermill site
Odell		
163 SP 966 577	2631	converted watermill
Old Warden		
164 TL 155 448	12131	watermill site
Pavenham		
165 SP 987 563	4427	windmill site
166 SP 989 552	8123	watermill site
Pertenhall		
167 TL 085 651	3160	windmill site
Podington		
168 SP 950 627	3159	windmill site
169 SP 928 646	3180	possible windmill site
Potton		
170 TL 213 498	0938	remains of tower mill
171 TL 234 501	10656	possible windmill site
172 —	10703	possible watermill site
Ravensden		
173 TL 064 537	3157	windmill site
174 TL 071 568	3158	windmill site
Renhold		
175 TL 092 509	0336	existing watermill building
176 TL 094 514	6606	windmill site
Riseley		
177 TL 047 625	0927	windmill site
178 TL 053 645	3025	site of smock mill
Roxton		
179 TL 149 550	1494	windmill site
Sandy		
180 TL 170 487	1111	watermill site
Sharnbrook		
181 SP 998 593	0929	converted tower mill
182 TL 011 591	1117	Stoke Mills (water)
183 SP 996 598	2632	site of Pinchmill (water)
184 TL 007 595	2830	horse wheel house
185 SP 999 595	7979	converted watermill

Parish & NGR	SMR	Notes
Shefford		
186 TL 149 395	0939	remains of tower mill
187 TL 147 395	2633	watermill site
188 TL 139 391	3156	windmill site
Shillington		
189 TL 119 356	1120	existing watermill building
190 TL 114 310	2625	site of Hexton Mill (water)
191 TL 138 355	3155	windmill site
Silsoe		
192 TL 070 357	3154	windmill site
Southill		
193 TL 171 409	2634	site of Stanford Mill (water)
194 TL 171 427	2828	steam mill building, Broom
Stagsden		
195 SP 983 476	3152	windmill site
Stanbridge		
196 SP 965 246	0535	windmill site
197 SP 962 245	0940	converted tower mill
Staploe and Duloe		
198 TL 110 601	3183	windmill site
199 TL 170 602	—	converted tower mill
Steppingley		
200 TL 011 360	6607	windmill site
Stevington		
201 SP 992 528	0924	preserved post mill
202 SP 983 546	1904	site of fulling mill (water)
203 SP 984 544	3556	windmill site
Stotfold		
204 TL 222 384	2611	Taylor's Mill (water)
205 TL 224 368	2612	Randall's Mill (water and steam)
Streatley		
206 —	3151	windmill site
207 TL 081 294	3197	windmill site
Studham		
208 TL 028 160	3150	windmill site
209 TL 008 168	10920	possible windmill site
Sundon		
210 TL 043 266	3149	windmill site
Sutton		
211 TL 213 485	3148	windmill site
212 TL 212 467	5133	possible windmill site
Swineshead		
213 —	6608	possible mill site
Tempsford		
214 TL 158 530	1137	water-powered saw mill
Thurleigh		
215 TL 048 582	0941	ruined tower mill
216 TL 056 587	7344	windmill site
217 TL 056 587	7345	windmill site
Tingrith		
218 TL 000 323	9508	possible windmill site

Parish & NGR	SMR	Notes
Toddington		
219 TL 022 294	1142	existing watermill building
220 TL 004 305	3143	windmill site
221 SP 989 285	3144	windmill site
222 TL 006 288	3145	windmill site
223 TL 016 293	3146	windmill site
224 —	3147	windmill site
225 TL 003 302	5213	watermill site
Totternhoe		
226 SP 990 202	0942	Doolittle Mill (wind and water)
227 SP 990 204	2826	windmill site
Turvey		
228 SP 938 524	2617	existing watermill building
229 SP 961 526	3142	windmill site
Westoning		
230 TL 040 322	3141	windmill site
Wilden		
231 TL 099 561	3139	windmill site
232 TL 107 556	11933	windmill site

Parish & NGR	SMR	Notes
Willington		
233 TL 118 505	0769+ 2635	watermill site
Wilstead		
234 TL 052 450	3138	possible windmill site
Woburn		
235 SP 961 332	1689	Park Farm watermill
236 SP 965 338	3136	windmill site
237 SP 941 329	8498	converted base of smock mill
Wootton		
238 —	6604	windmill site
Whipsnade		
239 TL 011 179	3140	windmill site
Wrestlingworth		
240 TL 254 476	3135	windmill site
Wymington		
241 SP 956 653	0931	windmill site